WHAT LEADERS ARE SAYING ABOUT MARK HECHT

"Mark Hecht is a master at his craft. He is passionate, wisdom filled, and uniquely qualified to share what matters most to great coaches-helping others become their best."

–Lynn Guerin, *Founder CEO,*
John R. Wooden Course

"Mark Hecht is the most talented leadership development coach I have ever worked with. He has global experience and a unique and simple approach to leadership development and coaching. Mark distills the essence of leadership performance down to a few basic questions which enable leaders to discover their own pathway to a stronger, more effective leader. I do not hesitate to call on Mark to engage on the most challenging leadership development opportunities. He challenges leaders to lead with vision, focus, and passion based on a strong foundation of core values."

–Mark Bogle, *Vice President,*
Acetyls Manufacturing and Tennessee Operations Site Leader,
Eastman Chemical Company

"I learned what real coaching is when I paused, looked back, and realized how much I've evolved and accomplished in a short period of time based on Mark's authentic coaching style. Forget all you know about coaching and disrupt the way you think about leadership. Mark leads you to find your own framework and skills for powerful, meaningful, and impactful business and personal relationships."

–Ana Costa, *Sales Manager,*
EMEA and Americas, Eastman Chemical Company, Portugal

"Mark's approach to developing coaches has a strong bent toward relationships with direct reports. His teaching style instilled in me the value of spending time with your employees to build trust and mutual ownership for the team's mission. The lessons he taught me have proved to be an asset for the last twenty-five years. Mark truly has a heart for managers who want to learn how to be effective leaders!"

–**Jeff Frazier**, *Training Manager,*
Tennessee Operations, Eastman Chemical Company

"Mark's wealth of experience and deep sensitivity make him a most successful coach and advisor. A master in positive transformation."

–**Michela Gambini**, *AMS Manager,*
Procurement, EMEA and Asia Pacific Regions, Eastman Chemical
Company, The Netherlands

"Mark's coaching programs have helped me to enhance my leadership potential and be recognized by management for the value I bring to the organization. For more than a decade I've been using his coaching to run the departments I lead, and it has never failed me."

–**Sabine Gehlen**, *Customer Service Manager,*
EMEA Region, Eastman Chemical Company, The Netherlands

"I've been in a leadership role now for 21 years and Mark taught my initial leadership class. I've also attended many of Mark's workshops over the years. He has been a huge influence throughout my career, and I continue to look to him for leadership guidance. Leading a successful team can be challenging but through Mark's superior teaching, mentorship, and influence I have been able to develop my abilities and confidence as a leader resulting in more effective, successful teams."

–**Sherry Good**, *Principal Team Manager,*
Technology, Eastman Chemical Company

"Either because of his personal qualities or because of his extensive experience and knowledge, Mark always seems one step ahead of us and therefore is a great source of guidance when it comes to coaching, mentoring, and building productive and fruitful teamwork."

–**Tatjana Ivanova**, *Site HR Manager,*
Eastman Chemical Company, Estonia and HR Business Partner for Russia

"During his career, Mark transitioned from a traditional HR role into one of truly helping others achieve their full potential in both their personal lives and their work. His ability to relate to people at all levels of the organization was truly unique and made him a much sought-after resource for a diverse set of functional groups within the company."

–**Rick Johnson**, *retired Senior Vice President,*
Eastman Chemical Company

"To his coaching practice, Mark brings his natural instincts, highly developed skills and years of commitment as a trusted advisor to all leadership levels in a Fortune 500 company."

–**Edna Kinner**, *retired Vice President,*
Human Resources, The Americas, Eastman Chemical Company

"Mark's extensive experience in coaching individuals at every level of an organization, across a broad range of functions and around the globe provides him with a truly unique perspective on the challenges associated with becoming an effective coach. The demonstrated success he has had in meeting this diverse set of leaders where they are at on their leadership journey and helping them transform their approach to coaching provides a level of insight that few have to offer. Translating this into pragmatic 30 daily readings is something I look forward to leveraging in my own leadership journey!"

–**Brad Lich**, *Executive Vice President & Chief Commercial Officer,*
Eastman Chemical Company

"I have been working with Mark for many years and I could not wish for another coach. In addition to understanding the disciplines of Management and Organizational behavior, Mark developed an acute sensitivity for the influence of culture on structure and processes across nations and cultures. This allows him to interact efficiently beyond the speaking of a language: You cannot coach leaders successfully in Estonia, Malaysia, or Tennessee without growing respect for other cultural perspectives."

–**Godefroy Motte**, *retired Executive,*
Eastman Chemical Company, France

"Mark Hecht and I started working together in the early 1990s, when he trained our first line supervisors and started informally being my executive coach in my role as Manufacturing Director. Mark is an exceptional trainer and coach on leadership competency and character. He cares deeply about people and treats everyone with respect and kindness, regardless of their company position. He is highly effective in developing leaders to be the best they can be. I finished my career 15 years ago and now work with Mark in the local non-profit arena and consider him a good friend and leadership coach to this day."

–**Roger Mowen**, *retired Senior Vice President,*
Eastman Chemical Company

"Mark Hecht has been an inspiration to me since I first met him in February 1999. His leadership models provided guidelines for me to become an effective leader. The idea of Helping Others Grow is a successful model that I have used to promote three of my direct reports to managerial level positions. He epitomizes what it means to be a coach to leaders."

–**Darrell Murphy**, *Principal Team Manager,*
Analytical Support, Eastman Chemical Company

"Coaching sessions with Mark were one of the most valuable, insightful and powerful transformational experiences both for me personally as a human being and for my leadership skills and career. Embracing and acknowledging your imperfections as a part of your unique personality and with the same respect as your strength, Mark helps you to gain the very best from the skills you have inside yourselves. Once this part is accomplished, you will start to see more and more positive changes in your own leadership style, tailoring your coaching process to fit individual learning needs and styles of people you lead. Mark's coaching was a thoroughly rewarding process that helped me to make massive strides towards a more fulfilling career and life."

–Alina Perederiy, *Customer Team Lead,*
Eastman Chemical Company, Germany

"I worked closely with Mark for over 20 years. He was more than respected, he was revered by management (internationally) for his character, competence, compassion, confidentiality, and ability to transform clients into more effective leaders."

–Ed Reynolds, *retired Senior Associate,*
Organization Development, Eastman Chemical Company

"Mark's training and coaching always brought to the table a human leadership, placing the limelight on listening, awareness, and the different values that motivate each person. 'There is nothing more unfair then treating everyone the same.' 'Do not focus on what you have to fix but on what you do best.' I will always remember these teachings. Mark is a master at the right questions so I could find my own answer. I do my best to practice the same approach with the teams I lead."

–Herminia Roman, *Sales Manager,*
EMEA Region, Eastman Chemical Company, Spain

"Having worked closely with Mark across the Asia Pacific Region over a period of 10 years, I have the privilege of seeing Mark applying his unique leadership and executive coaching skills, knowledge and experience to effectively improve and enhance the performance of leaders in the region. Mark brings a realistic and very practical approach to coaching."

–**NS Tan**, *retired Director, Human Resources, Asia Pacific Region, Eastman Chemical Company, Singapore*

"I wish everyone an excellent coach like Mark! He brings simplicity and practical insight to the coaching world."

–**Marian van Kempen**, *retired Director Human Resources EMEA Region, Eastman Chemical Company, The Netherlands*

"Would recommend attending one of Mark's coaching workshops! He is able to translate great theories and research into practical and concrete situations which allows the participant to learn more about leadership and coaching away from the beaten tracks. His workshops are a fantastic opportunity to be inspired and challenged on how you can make a difference in the lives of those you coach."

–**Matthijs Veenema**, *Global Business Conduct Director, EMEA Region, Eastman Chemical Company, The Netherlands*

"During my 30 years working experience, it was a great opportunity to meet Mark by getting his help and guidance to develop and extend my coaching and leadership skills as well as others in the organization. I should also say that he was the best in influencing across the region and organization, including non-business area."

–**Brian Yoon**, *Regional Business Director, Asia Pacific Region, Eastman Chemical Company, China and South Korea*

Revealing the Invisible

Coaching the People You Lead to Discover, Learn, and Grow

MARK HECHT

Published in the United States by Ignite Press.
ignitepress.us

978-1-950710-50-8 (Amazon Print)
978-1-950710-51-5 (IngramSpark) PAPERBACK
978-1-950710-59-1 (IngramSpark) HARDCOVER
978-1-950710-52-2 (Smashwords)

For bulk purchase and for booking, contact:
Mark Hecht
mark@coachinghorizons.net

Library of Congress Control Number: 2020911434

Cover design by Ravi
Edited by Dr. Beverly White
Interior design by Eswari Kamireddy

DEDICATION

To each member of my family, for making me a better person along my walk with you. You continually encourage me and bring me great joy. I love you far more than words can express.

To life mentors and friends, Dr. John and Mrs. Ellen Thompson.

To Tom Hornsby, Ed Reynolds, Rick Johnson, Roger Mowen, and Matt Stevens, along with the many Team Managers at Eastman Chemical Company, who gave me my start in a life of coaching.

To all those who allowed me over the years to impact their lives. What greater blessing is there than to hear someone say, "You made a difference in my life!" What I have learned along the way is that in impacting another it was always I who was blessed the most. Thanks to each of you, for you are my inspiration to write this book.

COACHING HORIZONS
Connect · Invest · Excel

(Painting by Amy Ferguson)

"The sunrise shall visit us from on high to give light to those who sit in darkness and in the shadow of death, to guide our feet into the way of peace."
Luke 1:78 (ESV)

CONTENTS

INTRODUCTION

In March 2016, I retired just shy of 35 years of service with the same company. Originally hired as an Eastman Kodak employee at the Tennessee Eastman site, I witnessed the spinoff of Eastman Chemical Company from the mother ship, Eastman Kodak, and was blessed to participate in the global growth of Eastman Chemical Company. After engaging in countless global coaching opportunities with leaders at all levels of the company and facilitating leadership workshops in multiple cultures and locations in over 20 countries across Asia, Europe, and the Americas, I will never forget the feedback from one leader at the end of a coaching workshop in the early years of my coaching work. It was a moment of feedback that will always remind me that the 25 years of global travel with 2.5 million flying miles was worth it:

> *"Thank you for the workshop and the power of coaching that we have experienced the last two days. I am now encouraged and hopeful that I can once again connect and communicate with my 17-year-old daughter whom I love dearly but with whom I have lost connection in the last couple of years."*

To be honest, I did not see that type of feedback coming. After all, this was a coaching workshop that was part of a broader corporate leadership curriculum. The purpose of the curriculum was to drive toward company goals, engrain corporate values, and create competitive advantage and global growth, all of them important to the success of a global organization. But the words from this leader caused me to pause and ponder what is key in reaching all those things important to the existence of a corporation or any other organization where people are working together to achieve goals and find success.

- What if a coaching workshop not only helped leaders work better through others to improve organizational success but also enabled them to impact others anywhere along life's journey in ways that very few in their lives will ever do?
- What if all leaders had a genuine interest in the people they lead, knowing their stories and knowing how every person brings value to the workplace?
- What if leaders were frequently having coaching conversations with others, helping others to see themselves in a different way and understand what keeps them from performing at a higher level?
- What is the impact of a coach who is curious in who you are, what you know, helps you discover your best talents and then expects you to use them, holding you accountable for bringing your best to what you do? How does that interest feel and impact the amount of discretionary effort you choose to give each day at work? How does it feel when you know someone expects your best every day? They refuse to let you settle into a life of mediocrity.

From that day till now, I have continually looked at coaching with a fresh perspective. The impact of those words many years ago from a workshop participant, along with the encouragement from others around the world who have allowed me to be a part of their lives, is the reason I have written this book. My hope for you is that the book will be an encouragement in your role of influencing the life of another, regardless of where you may be serving, and will enable you to be purposeful and intentional about the growth and success of those who have been entrusted to you.

Coaching is an investment in others based on the commitment to see them through success and failure, the highs and the lows. We never give up. We stay the course. We are committed to their growth. We stay with them because we are committed to who they are and to who they are becoming. Coaches have a bias, a belief that talent is there to be found in people. Coaches sometimes see in people what others cannot see or refuse to see. People have amazing capabilities, and often many of our capabilities stay hidden within us.

People are like the white light before it enters a prism. What does a prism do? A prism shows us the primary colors that are naturally in white light but that we cannot see without the white light going through the prism. The invisible which we cannot see becomes visible because of the prism. The power of coaching is like the power of a prism; and in the metaphor of the prism, we can find the essence of what coaching another is hoping to accomplish–revealing the invisible. From there, coaches help others move from discovery to learning to growth on the journey to becoming the best version of themselves at work and sometimes impacting other areas of life that also need them to be at their best.

> **"***Coaches sometimes see in people what others cannot see or refuse to see.***"**

What's in it for you, the leader, to be willing to make such a commitment to others? In this book, you will hear the clear message that, as the leader-coach, the coaching of others is not about you. To invest in another person is to make them a priority; they grow as you move behind the curtain. They get the credit while you take the heat standing beside them, or sometimes even in their place. With that said, hopefully life has already shown you the following to be true; *"Invest in others, and like a boomerang, it will come back to you, sometimes in a most unexpected way. When people genuinely desire to add value to others, they cannot help others without receiving benefit."*[1] Often when we give to others what we would most like to receive, it comes back. As we trust others, we become more trusted. As we help others, they look for ways to help us. As a leader-coach within an organization, we want to see the organization benefit; and coaching provides an avenue to increase job satisfaction, motivation, and excellence, which makes any work environment a better place to be and, on the business side, contributes greatly to competitive advantage.

Within the 30 daily readings, I will share reflections from my journey in coaching others and from what I have learned engaging with leaders at many levels within various organizations as they have coached others. I am very thankful for the leaders over the past 30 years who have allowed me to walk with them in their leadership journey. The reflections target what I have seen makes a good coach and what can elevate a good coach to be even better. Some days focus on foundational coaching concepts. Other days focus on specific coaching skills which the best coaches have taught me really make a difference in coaching effectiveness. At times I toggle between reflections on coaching and leadership. Some days you will see me lean somewhat toward the philosophical, with reflections on coaching mindsets regarding topics like failure, busyness, perspective, and time. If we fail to pause and ponder these deeper mindsets, we will fall short in making a true difference in the lives of anyone.

Throughout the book, I share many quotes by those who say things better than I could ever say them. In the Endnotes, you will find references to resources and experts who have helped me on my coaching journey. Follow these resources from your own interest and curiosity. We all benefit from standing on the shoulders of others, their innovation and courage to step out and change the world, even if just one person at a time.

The following teaching has often encouraged me as a coach:

"Stay alert, stand firm, be courageous, be strong. Do everything in love."
1 Corinthians 16:13-14 (ESV)

How can this teaching help you have more impact and be more consistent in coaching others?

- *Stay alert,* be attentive to what others need from you to grow.
- *Stand firm* in your belief that every person has value to add and the potential to grow. There is the invisible within, waiting to be discovered.
- *Be courageous* enough to bring a tension, a push, and the honesty

needed to help others make progress on their best path forward. When done within an organization, coaching links their growth to that which best impacts the organization.

- *Be strong* to stay the course with each person you coach, never wavering, always expecting their best.
- *Do everything in love.* What does each person you coach need to know about you? That you care deeply for them as individuals, and that you know how the work they do adds value. Also, frequent acts of kindness toward them always sends a powerful message.

May the 30 days of reflections help you grow as a coach and enable you to experience even greater success in revealing the invisible within those you lead. And may you discover that as you bless others you too may obtain a blessing.

DAY 1

THE DIFFERENCE A COACH MAKES

"When you hire a hand, it comes
with a head and heart attached."

PETER DRUCKER, MANAGEMENT CONSULTANT,
EDUCATOR, AUTHOR[1]

Over the years, I have often asked my coaching class participants this question: "What is the impact of a person's immediate supervisor on how much discretionary effort they choose to give on the job each day?" The answers often range from 20% to 80%. Those with a history of good bosses tend to underestimate the value of a good boss and state a lower percentage of impact. For those who have experienced a period of working for a poor boss, the number soars, some saying the impact on both work and family life can be miserable.

A 2015 Gallup study found that managers account for at least 70% of the variance in employee engagement, how much of themselves they bring to work.[2] As you ponder your work experience, you will have to make your own conclusions on the percent of impact; but for sure, every day as people make choices on how much effort they will give at work the immediate supervisor has a significant impact on that choice!

What is discretionary effort? Can you recall a workday when you chose just to get through the day? You give good effort, but the focus seems to stay on the things that "have to" be done and then you are glad to call it a day. How did you feel at the end of that day? Glad when it was over? Whatever the reasons for the choice, when we approach our workday from the choice of "have to," we call that compliance. It is not bad or just average work, but we know inside that on those days we did not bring our best to our work.

On other days, we make a different choice. We bring more of ourselves to the work: more energy, creativity, thought, perspective, and even passion. When we approach our workday from the choice of "want to," we call that commitment. Every day we make a choice. This choice between compliance and commitment is called discretionary effort.

A class exercise I do at times with leaders starts with asking them to raise their hand. Each leader will casually look around the room to make sure others heard the same request. Eventually everyone will half heartily raise their hand with their elbows bent. Then I ask them to raise their hand as high as they can. Up the hands go with the arms very straight. That difference between a partially-raised hand and one fully-raised is a picture of discretionary effort, the choice of bringing more of oneself to one's work. In that choice lie job satisfaction, motivation, excellence and competitive advantage. That choice is significantly impacted by the relationship, interactions, conversations, and feelings toward my supervisor. If you are leading others today, do you realize that there is an emotion attached in working for you? What feelings do you think others have when they experience working for you?

> "*Every day we make a choice.* "

The question "Why do people work?" has been the topic of discussion in many different forums across the decades. I learned after teaching and coaching in over 20 countries that there is not one simple answer to that question. Just like each person we meet has a unique life story, each person works and finds in work something different. What I do believe is that most people want to do well in their work and each person makes a choice what effort they will give each day. We all have lives that fill our heads with responsibilities, joy, worries, hope, disappointments, success, failures; and each day we seek to balance those thoughts in our head with the tasks that we need to get done on the job today.

My hope is that this book of reflections on coaching will help you be more intentional about the work environment you create as a leader and help you better understand how the environment you create impacts the amount of discretionary effort that those you lead and influence bring to their work day. Leading with the intent and purpose to coach well will make you more consistent and greatly influence the degree that people will "raise their hands straight up" when they approach their work.

As coaches, you can choose to make a difference.

"Treat your peers as interesting fellow humans, and you may be surprised what it does for their motivation, dedication, and engagement."

CAMILLE FOURNIER, CTO, SPEAKER, ENTREPRENEUR[3]

DAY 2
GIVING HOPE

"I saw an angel in the stone and carved to set it free."

MICHELANGELO, ITALIAN SCULPTOR, PAINTER, ARCHITECT, POET
OF THE 1500S[1]

In the summer of 2016, five months after my retirement, I went back to Gallup headquarters in Omaha, Nebraska, to renew my 12-year certification in StrengthsFinder, my favorite self-assessment. It is my favorite because it focuses on how we excel by challenging us to make our talents our everyday strengths. At one of the breakout discussions during certification we were asked to define coaching. One of the participants at my table went to the flip chart and drew a prism to picture his perspective of coaching; at that moment, I fell in love with a new metaphor on what coaching can achieve with people. At the same time, I discovered the logo for my new business, Coaching Horizons.

"We each do something better than anything else we do, and sometimes we do certain things better than anyone else around us."

As noted in the Introduction, a prism shows us the primary colors that are naturally in white light but that we cannot see without the white light going through the prism. The invisible which we cannot see becomes visible because of the prism. In the metaphor of the prism, we can find the essence of what coaching another is hoping to accomplish–revealing the invisible within. People possess talents. We each do something better than anything else we do, and sometimes we do certain things better than anyone else around us. People are like the white light before it enters a prism. Coaching serves as a prism to bring out the talent, helping others to discover what is already within them. Our desire is to draw out and encourage them to be their best.

As a coach, do you believe that people are talented, have valuable contributions to make, and can grow to a higher level of contribution? I believe these assumptions regarding people are foundational to the effectiveness of any coach. Please note what I am not saying. I am not saying that everyone is at the same level of talent or capability. We all share inherent dignity, worth, and value because we are created on purpose and with a purpose. Yet, we are also uniquely designed in our talents and cognitive abilities. It was once explained to me that the talents of people we coach can be compared to a skyscraper scene.[2] Some are on the 40th floor, some are on the 50th floor, and others are on the 100th floor. We cannot move everyone to the same floor of talent, but we can challenge each person to move up additional floors from where they are today. We can help others see themselves differently and hold them accountable to be the best version of themselves.

C.S. Lewis summarizes it well when he said, "*We are not yet even half-done.*"[3] People have amazing capabilities, and often many of our capabilities stay hidden within us. Coaching can help bring out the unique design within others so they can uniquely impact their world.

There is a popular story regarding Thomas Edison, often described as America's greatest inventor, that may inspire you to look for the invisible within the people you coach.

When Edison was a young boy, a note was sent home from school to his mother that only she was to read. When young Thomas asked what the note said, his mother tearfully responded to him, "Your son is a genius. This school is too small for him and does not have enough good teachers to train him. Please teach him yourself." Years later after his mother died, Edison found the note as he went through her belongings. He was stunned and wept for hours after reading it. What the note said was, "Your son is addled (mentally deficient, unable to think clearly). We will not let him come to school anymore."

Edison later wrote, "My mother was the making of me. She was always so proud of me and I felt I had someone to live for, someone I must not disappoint." An entry in his diary says, "Thomas Edison was an addled (mentally deficient) child who by a hero mother, became the genius of the century."[4]

While the story above may be part truth, part legend, Edison's journal entry clearly shows that his mother saw within her son what others did not or chose not to see. She believed there was within her son something that needed to be revealed.

Do you coach with the hope of being a prism to the talents within another person? You will be a source of encouragement to those you coach, for giving hope to others lies in what you believe about them long before what you and others are able to see.

"A coach's primary vision must be focused not on seeing limitations but on seeing possibilities."

JOHN WOODEN, UCLA LEGENDARY COACH, WINNER OF 10 NCAA NATIONAL CHAMPIONSHIPS[5]

"Promise me you will always remember,
You are braver than you believe,
and stronger than you seem,
and smarter than you think."

CHRISTOPHER ROBIN TO WINNIE-THE-POOH[6]

DAY 3
CURIOSITY IN CONVERSATION

"We keep moving forward, opening new doors, and doing things because we're curious and curiosity keeps leading us down new paths."

Walt Disney, Pioneer of Cartoon Films,
including Mickey Mouse[1]

While the years have taught me that coaching well is not easy, in recent years I have learned that a powerful key to coaching others well is found in the simplicity of something we do every day, the simplicity of a conversation. Coaching happens between people and involves the use of words. Words create worlds. Words create change. Words shape behavior. Coaching is a conversation. Connect those two words forevermore–coaching and conversation. You can use your life experience with both coaching and conversations to become a better coach and help others move to a better place. All of us most likely can think of a person, a coach, who has impacted our lives. We remember what that person said and did to move us to a better place. We remember how that person made us feel. Also, we all have conversations every day. There lies the foundation for your coaching, drawing on experiences already in your life–coaching and conversations! Yes, there is uniqueness about a coaching conversation and what it's trying to achieve with another person, and we will reflect on that uniqueness. But always keep before you the simplicity of conversation as well. Coaching is a conversation.

A second learning that is equally key in coaching others well is curiosity in conversation. Coaching with curiosity is motivated by genuine interest in what the person is thinking and already knows. Seeking what the person knows and is thinking is priority over telling the person what you think and know! Curiosity withholds judgment, suspends your agenda, avoids fixing the person, and resists the temptation to co-own or take responsibility for the solution.

When was the last time you were truly curious in a conversation with another person? In our world with demands throughout the day, one might ask, "When do I have the time to be curious in conversation?" The demands and expectations put on leaders in our fast-changing work environment move most of us to operate in a cycle of *examine the problem, get the answer, implement the solution.* The idea of curiosity in conversation initially conjures up the idea of a long conversation with delayed answers and slow implementation of solutions. In day-to-day practice, however, mastering the art of curiosity in conversations can bring better answers, more creative solutions, and timely results.

I once sat in on a coaching class taken by a team of senior leaders. The leaders were encouraged to step back from directing the conversation to focusing on drawing from the other person their ideas and perspectives. During the early part of the class, the leaders evidenced some discomfort in the time spent to have such a conversation. But as the day went on, and they had coaching discussions around real issues, you could see a shift in the value they were placing on having a real conversation. During a specific coaching session, one of the senior leaders commented that she and her fellow team member had been discussing the same topic for months, but not until the conversation that day did she hear her partner share certain perspectives, concerns, and ideas that in one hour helped them make more progress than they had made in months of discussions on the topic. What was the difference? They had both experienced moving from the mindset of diagnostic coaching to curiosity in conversation, resulting in the discovery of new paths.

So how do we get to curiosity in conversations? To get there implies that we must come from somewhere else. And that is where the challenge lies. Leaders are smart. They are smart, whether at the 1st level supervisory position or at the senior executive level. Having the subject matter expertise in their field has brought them an abundance of recognition and rewards during their career. Examining the problem, finding the answers, fixing things has been the foundation of their careers long before they took a leadership role. Diagnostic habits are engrained and have served them and the world around them well. Diagnostic conversations have as their motive the collection of more information so that better advice can be given to fix the problem. Giving advice is not evil, but advice may be harmful to others who are learning and gaining insights as they work to figure out a situation on their own. To get to curiosity in conversations, we must move away from diagnostic conversations.

Try it. Coaching in its very essence is simply a conversation, and most things of value in life come from a conversation with another person. Curiosity in the coaching conversation keeps the learning, thinking, responsibility, and action with the person being coached. It initially will feel like strengthening a new muscle, and it takes a step of courage, for moving to curiosity brings uncertainty, a sense of losing control and the giving of trust to another person.

> **"To get to curiosity in conversations, we must move away from diagnostic conversations."**

The next time you are in a coaching conversation, be curious and seek out:

- What does the person already know about the challenge?
- What is being assumed?
- Where is the focus or what is the person paying attention to?
- Is the focus helping or serving as a barrier?
- What is the person hoping to accomplish?

Then watch how the power of your curiosity in conversation helps the person discover, learn, and grow toward a better place.

"Curiosity is more important than knowledge."

ALBERT EINSTEIN, GERMAN-BORN PHYSICIST, DEVELOPER OF THE THEORY OF RELATIVITY[2]

"I have no special talents. I am only passionately curious."

ALBERT EINSTEIN[3]

DAY 4

BE HERE NOW

"A coach listens not because listening is a good technique, effective, or because people like to be listened to. A coach listens because he believes the person has something to say."

TONY STOLTZFUS[1]

I remember hearing the story of a missionary, blind since a young boy, who shared his experience after surgery that gave him his vision back after 45 years of living in blindness: *"Having the physical capacity of sight is not the same as seeing. The visual input was very distracting when listening to others because it interfered with the way I had listened to others for 45 years, without sight. I had to learn to die as a blind person before I could live as a person with sight."*[2]

The story reminds me of the challenge we have in being present with others in conversation. Being physically present with others is not the same as having presence with others. We might be physically present but not "here now." If coaching in its very essence is a conversation, growing in our ability to have presence with others in conversations is key to our success as a coach.

Presence can be defined in many ways. The ability to fully experience what is around you at the very moment it is happening. Living in and appreciating the moment. Being where you are, now. My friends in China taught me the Chinese symbol for presence, which means "undivided attention." The symbol includes the ears, eyes, and heart all playing a part in giving someone undivided attention. How you view the person is also important, and the key is to give the person the value and respect of being the most important person in the world at that time. We are created as relational beings and long for someone to look us in the eyes and know who we are as individuals. As a result, rarely will we approach presence with a person if we do not value that person.

A challenge in keeping presence with others lies within how our brain works—we think faster than someone can speak. Most people speak at a rate of about 100 words per minute; however, we have the capacity to listen to someone speak at a rate of about 400 words.[3] There lies a key to presence. What does your mind do with the 300-word gap when you are in conversation with another person? Do you have any idea?

If you pause a moment and think about it, you know already what you normally do with the 300-word gap; your mind wanders and often creates a conversation within your head, losing focus on the conversation with the person. For me, my conversation within can easily move to figuring out how to solve the problem I am hearing from the other person. I fall in the trap of moving to diagnostic thinking instead of staying curious. Where does your mind go, and what is the conversation within that causes you to lose presence?

> "We are created as relational beings and long for someone to look us in the eyes and know who we are as individuals."

When we move to the conversation within, our brain brings another challenge in keeping presence. Now we are hearing words from the person with whom we are having a conversation while we are listening to the words from the conversation within our head. We think we can solve this challenge by asking the brain to multitask, listen to both conversations; but multitask the brain does not do![4] Actually, rather than multitask, when we attempt to listen to both conversations at the same time, the brain switches back-and-forth between the two tasks, like a toggle switch. This design of the brain does not hinder us when we are performing different types of activities (I can listen to you and change a light bulb), but when both activities include the need to listen to words, we do not hear either conversation well nor do we maintain presence with the other person.

Try it. Try listening to a television program while reading a magazine. You will miss content from both. Try reading a magazine and listening to your spouse, significant other, or friend speaking to you. You will again miss content from both, but now you are in trouble for ignoring someone important to you!

"Wherever you are, be all there."

JIM ELLIOT, MISSIONARY, MARTYRED WHILE SERVING ON THE
MISSION FIELD, 1956[5]

If you struggle with staying in presence with others, take the time to increase your awareness of when you move to your conversation within. John Whitmore reminds us, *"I am able to control only that of which I am aware. That of which I am unaware controls me."*[6] Don't try to fix it; simply increase your awareness of when you move and lose presence. Over time, awareness alone will increase your presence with others, for you will gain better awareness of when you move away from presence, enabling you then to come back. As you grow in presence with others, you may still find it challenging with certain individuals. That is a reality in life, as sometimes we don't easily connect with a person for various reasons. The solution? The best chance of improving presence in these situations is to look, look hard if necessary, for something unique about the person. Knowing their story also helps.

Do we dare mention the impact of the smart phone and laptop on our presence with others? Are you aware of when or how many times you reach for your smart phone when in a conversation with others or in a team meeting? How often does your open laptop screen take away your presence in a team meeting? Electronic multi-tasking, or what we think is multi-tasking (remember, the brain does not multi-task but toggles between one activity and another, never present in either one), has the same limits that we reflected on above; when two activities include the need to listen to (or read) words, we do not process the two activities well nor do we maintain presence with the other person

Fortunately, it is much more frequent today for teams to set ground rules for the non-use of electronic devices in meetings, and especially now with the increased use of virtual meetings, for we all know how tempting it is to want to "toggle" between activities when our microphone is on mute! For leaders, when in a meeting with people below you in the organizational hierarchy, always avoid the use of electronic devices, for it will almost 100% of the time be interpreted as "You have left the room." And the message received, loud and clear, by others from you is, "We nor the topic is important to you." Better to step out of the room if you need to take a phone call, and then come back to the room and maintain presence with those in the meeting.

One last thing I will mention here that impacts our ability to stay in presence with others is our bias for action. As leaders, coaches, and in the other roles we play in life, we are busy; and busyness causes us to lose presence with others. A study of goalkeepers in football (soccer) defending against penalty kicks is applicable to presence with others:[7]

- Goalkeepers who dive to the right to stop a penalty kick stop the ball 12.6% of the time.
- Goalkeepers who dive to the left to stop a penalty kick stop the ball 14.2% of the time.
- Goalkeepers who stay at the center have a 33.3% chance of stopping the ball.

How often do goalkeepers stay in the center? Only 6.35% of the time! Why? Our bias for action. It looks and feels better to miss the ball by moving, demonstrating action, even if diving in the wrong direction, then to have people wonder why you, the goalkeeper, just stood there and watched the ball go by! Are we any different in our world of busyness than a goalkeeper? Stepping back in our conversations with others, staying curious longer, delaying fixing, and actually listening to what people say to us—these are the fruits of presence.

Few gifts in life send the message of value and respect to others than the gift of being in presence with them. It eliminates the feeling in others of "I am just a transaction," giving others instead the feeling of connection. People long to know and be known.

WARNING: As you grow in presence with others you might be shocked to learn how much you may have been missing in life!

"Trust is earned in the smallest of moments. It is earned not through heroic deeds, or even highly visible action, but through paying attention, listening, and gestures of genuine care and connection."

BRENÉ BROWN, RESEARCHER, STORYTELLER[8]

DAY 5

FARM (FOCUS)

"The focused mind only picks up on those aspects of a situation that are needed to accomplish the task at hand, the here and now. It is not distracted by other thoughts or external events."

TIM GALLWEY, FOUNDER OF THE INNER GAME[1]

Focus on the **+** in the picture below. Keep focusing. What happens?

Hopefully, you can see that as you focus on the **+** and keep focusing, the outer circles fade or almost disappear. So what does this have to do with coaching?

Let's step back a moment and answer the question, "What is a coach actually attempting to do in the coaching conversation with another person?" We have reflected on the power of curiosity in conversation, but what is the product or fruit of the coaching? Since coaches focus on helping people grow, I like to summarize the product of coaching in the acronym FARM. A coach steps into coaching conversations with others for the purpose of:

Sharpening **F**ocus
Increasing **A**wareness
Creating **R**esponsibility
Facilitating **M**obility

Today, let's reflect on focus. I like to use a skiing analogy to create a word picture on the importance of focus, as well as awareness, which we will reflect on in Day 6.

I am told by skiers that the rule of skiing is "where you focus is where you go." When leaving the top of a mountain, what is the focus and hope of the skier? The bottom of the slope. The skier wants to be aware of the obstacles (analogous to the outer circles on the picture above) going down the slope, but the focused goal is the bottom of the slope. When I was sharing this word picture in class a few years back, a motorcycle biker told me the same rule applies to riding a motorcycle. When the rider on a motorcycle loses traction on the road, what is the only hope for staying out of the ditch? Focusing back on the road in hopes that the motorcycle will follow.

Coaching follows the same rule: *Where the person focuses is where they go.*

And the quality of their focus lies in your mastery of the coaching conversation. The coaching conversation is the vehicle that helps the person focus. We stay curious in conversation and seek what the person is focusing on:

- What is the person paying attention to?
- What are you hearing as they share their situation?
- What topic do they keep repeating?
- Where are they stuck?

Focus is paying attention. When people focus on something, they are paying attention to it. It is in the coaching conversation where we discover where the focus is and then help the person determine if a change of focus or a sharpening of focus is needed.

"Where you focus is where you go."

Focus is the defining difference in performance. The most successful athletes and athletic coaches have learned the link between focus and performance. It is no less of a reality in business and performance coaching. Focus and awareness go hand and hand. Let's move to Day 6.

"People think focus means saying yes to the thing you've got to focus on. But that's not what it means at all. It means saying no to the hundred other good ideas that there are."

STEVE JOBS, CO-FOUNDER OF APPLE COMPUTER[2]

DAY 6

FARM (AWARENESS & RESPONSIBILITY)

"Until we help a person increase awareness of the gap between what he thinks he is doing and what he is actually doing, he cannot move to correction and improvement."

PETER DRUCKER[1]

Let's revisit the skiing word picture. As a skier, in addition to knowing where the bottom of the slope is, my focused goal and destination, I need to be aware of other things if I am to ski down the slope successfully. To start with, where do I start my journey? Also, as I start skiing, what might hinder my ability to get down the slope? In skiing, there are obstacles that I must be aware of or I can get hurt-trees, rocks, and other skiers (analogous to the outer circles on our picture from Day 5) to name a few.

As the coaching conversation helps to sharpen focus, the person we are coaching will begin to become more aware of the following:

- Where am I now?
- Where do I want (or need) to go? (focused goal)
- What do I need to do differently, at least first steps, to get there?
- What might hinder or get in my way?

John Whitmore wrote, *"Awareness is knowing what is happening around you. Self-awareness is knowing what you are experiencing."*[2] Awareness is a journey of learning, of gaining new insights. Learning what is within and what is around us impacts our choices and behavior. Without increased awareness, the coaching conversation will not be fruitful. This awareness need not be perfected at the beginning, but the message we want to send upfront to the person we are coaching is, "Without a desire on your part to pause, think, and learn, this conversation will not be helpful to you." Underlying this is the strong message of responsibility. From the beginning to the end, the person being coached must take responsibility for engaging in the coaching conversation and is responsible for doing something with the sharpened focus and increased awareness.

Coaching creates responsibility in others, which is why we approach the coaching conversation with curiosity rather than with diagnostic habits. If we, the coach, fix or solve everything, then we also keep the responsibility for everything. As we reflected on Day 3, it is so easy for the coaching conversation to become about the coach's need for awareness, more information, more data and details so they can quickly move to fixing. We will reflect on this much more in upcoming days, but the person being coached is who needs to think and learn through the coaching conversation, not you. Your role is to help facilitate the learning that leads to sharpened focus and increased awareness.

> **"*Without increased awareness, the coaching conversation will not be fruitful.*"**

With focused attention comes awareness and clarity on where to go. While it is important to be aware of the obstacles that could hinder performance, with sharper focus and new awareness, the obstacles begin to fade away and improved performance is now a more likely outcome.

Where a person is focused is where they go, and the increased awareness heightens the focus. Focus and awareness build upon each other, increasing the likelihood of successful mobility toward the goal.

"Awareness is about knowing the present situation with clarity."

TIM GALLWEY[3]

DAY 7

FARM (MOBILITY)

Christopher Robin:	I wonder which way.
Winnie-the-Pooh:	I always get to where I'm going by walking away from where I have been.[1]

Coaching never leaves a person in the same place.

Please read the above sentence again. Coaching is a conversation, but it is unique when compared to other conversations in that it is always intentional, focused on moving a person from where they are to another place. All coaching conversations end in action. If the person you are coaching does not have a next action step in mind at the end of the conversation, is not moving to a better place, you have not had a coaching conversation. You may have had a helpful conversation in some other way, but it was not a coaching conversation.

Coaching is mobility, momentum, action, movement toward a focused goal.

> **"Coaching never leaves a person in the same place."**

A step back into the history of the word "coach" can help us on this perspective. The English word "coach" first came into use in 15th century Europe, derived from the Hungarian word "Kocis" (carriages), after the village of Kocs, where transportation coaches were constructed.[2] What is the purpose of a carriage, a coach? To transport a person from where they are to where they want to go. Likewise, in your role as a coach, you are helping another person move from where they are to where they want to go. Stamp the picture of a carriage (coach) in your mind to remind you that when you are coaching you are *"facilitating the mobility of another."*[3] They may not always know exactly where they want to go, and some may not know where they are yet, but in coaching you are helping them move.

Mobility completes our FARM acronym. Coaching is simply a conversation, but it is also unique in what it enables others to do. Coaching is not counseling or therapy; the focus is different. There is a time and process for healing, relieving pain and hurt, or dealing with deep issues from the past, but coaching is helping others gain focused attention on the things that make a difference in performance today. Coaching increases awareness of what is within and around the person and then lays before them the opportunity to take responsibility to move to a better place.

A coach steps into coaching conversations with others for the purpose of:

Sharpening **F**ocus
Increasing **A**wareness
Creating **R**esponsibility
Facilitating **M**obility

As we come to the end of our reflections on FARM, I leave you with my favorite statement on the essence of coaching:

> *"Coaching is the genuine interest in another person, paired with curiosity in conversation, which enables the person to learn what is within and perform at their best with new awareness and focus."*[4]

As you coach others to discover, learn, and grow, may FARM, representing the product of the coaching conversation, enable you to be more successful.

"The way to get started is to quit talking and begin doing."

WALT DISNEY[5]

DAY 8
FAILURE AND DISAPPOINTMENTS

"Don't permit fear of failure to prevent effort. We are all imperfect and
will fail on occasions, but fear of failure is the greatest failure of all."

JOHN WOODEN[1]

I would like to pause on this day to be very clear about the intent of this
book. Brené Brown in her book, *Dare to Lead*, reminds us that when
giving feedback, "Clear is kind, to be unclear is unkind."[2] So, let me be
clear, and you can decide if I am being kind!

This book is intended to encourage you as a coach to impact others
and the worlds they serve, to be more intentional and purposeful in your
coaching, thus being more consistent and predictable, which builds their
trust in you. To be clear, this book is not a judgment or critique of how
and when you will fail in your role as a coach of others. Being people, we
will fail at times.

Two things I have found are certain about coaching. First, it is like
math. You don't get good at math unless you do the problems. Likewise,
you will not get better at coaching unless you coach. Second, when you
coach today, you are coaching better than you did yesterday but not as
good as you will coach tomorrow. Coaching is a journey.

"We are people in relationship with other people,
and all those relationships bring disorder and
unpredictability."

I would have to add several days of reflections to this book to cover all my failures as a coach, but I will share a few with you here:

One early December I walked past the office of a leader with whom I had been meaning to provide some feedback. I distinctly remember telling myself, "We have a good relationship; let me go ahead while he is here in town and share the feedback with him." I then proceeded to go into his office, share the feedback, and ruin his Christmas holidays! From that day forward, I have not given feedback to another without more thoughtful reflection, preparation, and confirming the right time to give it.

A senior leader spent $18,000 on a ticket to fly me across the world over a four-day period to facilitate a conflict resolution session. It ended up being a $18,000 investment in teaching me how *not* to facilitate a conflict resolution session! What most impressed me from that failure, however, was how that leader never held that failure against me. He truly saw the failure as an investment in me, expected me to grow from it, and never hesitated to involve me in future challenging coaching situations.

I found myself caught between a team crying out for help via confidential channels and the leader of that team, a friend of mine, whom they saw as the source of their pain. At the end of the journey, I believe the team ended up in a better place; but a few years later, when the leader discovered that I had played a role behind the scenes in the situation, he was quite hurt. I am not sure how I would have handled it differently if given the opportunity again; maybe there was a better path. It was the timeless challenge of balancing the good of one with the good of the many. How do you protect the many while honoring the one? How do you protect the one while honoring the many? I am not about to claim I have solved that challenge. But for certain, it was a learning experience for me to face the leader and listen to how the situation had harmed him and to be reminded that in the work of coaching and leading others, sometimes things turn out nicely for all, and sometimes, even with good intentions, someone gets hurt. I became a better coach after that experience and learned to say more quickly, "I am sorry," when my actions harm another. I also witnessed courage as I know it was not easy for the leader to confront me and communicate the harm that he had experienced.

One more. I once received some very candid negative feedback from a peer. It was not delivered in a skillful manner, making it initially more hurtful than helpful. After getting through the initial hurt, I needed to decide if I would use it to get bitter or better, for hidden behind the method of delivery was truth which I needed to hear. Once I decided to use the feedback to get better, I was able to ask the question, "What had I done to make him see me the way that he did?" It was a reminder that at times we can so easily think too highly of ourselves. I had failed myself in not keeping a spirit of humility; and, fortunately, while the feedback was not delivered in the best way, I benefited from hearing what needed to be said.

I share these stories to illustrate that there is no intent or hope of this book being written in perfection or that the reading of this book will make you a perfect coach. We walk through, live in, and are surrounded by failure. You have and will have your stories of failure and disappointments. We are people in relationship with other people, and all those relationships bring disorder and unpredictability. We will disappoint each other.

"Success consists of going from failure to failure without loss of enthusiasm."

WINSTON CHURCHILL[3]

DAY 9

THE POWER TO RECOGNIZE, ADMIT, LEARN, AND FORGET

"Failure is valuable if you do four things with it: recognize it, admit it, learn from it, forget it."

JOHN WOODEN[1]

Those we coach will need help and encouragement to engage with disruptions in their lives, their failures and disappointments. While failure is essential to learning, feeling like a failure adds no value and can stop us from moving forward. If the feeling grows, it becomes like barnacles on the bottom of a boat, stopping the smooth flow across the water. The feeling of failure is like a rip in your favorite piece of clothing.[2] If not repaired, the clothing will be ruined. The feeling of failure takes away opportunities for joy and satisfaction in the moment of learning.

The quote from Coach John Wooden at the beginning of today's reflection can be very helpful in guiding a person through failure. First, it is important to recognize failure and to admit the part we played in the failure. Recognizing that failure has happened and that one's action, decision, or plan was a contributor to the failure are important first steps. Recognize and Admit. Like awareness, unless our perception matches reality, we cannot progress forward.

Next, how can we help the person learn from the failure? Learning avoids further ripping in the fabric.

"If you win every time, you don't learn anything. You don't learn anything about yourself. You don't learn anything about the other person. You don't learn anything about the game. You don't learn anything about life. To get beat is very healthy, especially when you have given it your best effort."

JACK NICKLAUS, WINNER OF 18 PGA MAJOR CHAMPIONSHIPS[3]

In conversation, give the person a safe place, not only to recognize and admit but also to share what they learned from the situation that they can take forward. Keep the conversation focused on learning; because, without learning, the person will only repeat history.

The last step: Forget it. This initially sounds contrary to the learning step; but in the context of performance, it is critical. To forget is to keep the failure from becoming a source of interference for future performance.

"I didn't leave her there (the bench) *for long. When a player makes a mistake, you always want to put them back in* (the game) *quickly – you don't just berate them and sit them down with no chance for redemption."*

PAT SUMMITT
UNIVERSITY OF TENNESSEE LEGENDARY COACH, WINNER OF 8
NCAA NATIONAL CHAMPIONSHIPS[4]

"After failure, learn. Then delete. The great athletes push the delete button after every bad play. The bad play does not stay in their heads."

INNER GAME ARTICLE

"I have missed more than 9000 shots in my career. I have lost almost 300 games. 26 times I have been trusted to take the game winning shot and missed. I have failed over and over and over again in my life. That is why I succeed."

MICHAEL JORDAN, NBA BASKETBALL HALL OF FAMER[5]

A key here is how you, as the coach, carry the failure into the future. Recall my failed experience I shared on Day 8; my leader did not hold a $18,000 failure against me but used the investment he had made in me. The coach plays a critical role in a person forgetting and moving forward. At the same time, I recognize and have been a part of failure where the risk and impact of the failure was too broad to only learn from it and forget it. The failure was outside the reasonable risk zone, and consequences such as loss of job or change in role were necessary. That is a reality of life, but many, if not the majority of, failures lie within the reasonable risk zone, and it is within this zone where learning can take place. Look for opportunities to coach people through the process of recognize it, admit it, learn it, and forget it, a powerful process for coaching others through failure.

> **"While failure is essential to learning, feeling like a failure adds no value and can stop us from moving forward."**

When the person you are coaching gets stuck and has difficulty moving through the process of failure, there may be several root causes. But often you may find one of two things at play: their definition of success and the degree in which they compare themselves to others.

First, ask yourself and be curious to know, "How is the person defining success?" Is their definition of success making it hard for them to admit or learn or forget? Two perspectives on success for you to ponder, and maybe challenge others with:

"Success is peace of mind attained only through self-satisfaction in knowing you made the effort to do the best of which you're capable."

JOHN WOODEN, THE PYRAMID OF SUCCESS[6]

"If I had to embrace a definition of success, it would be that success is making the best choices we can…and accepting them."

SHERYL SANDBERG, CHIEF OPERATING OFFICER OF FACEBOOK[7]

Second, are comparisons to others becoming a barrier to moving forward? Comparisons seldom set us up to honor who we are.

"Never try to be better than someone else. But always be learning from others. Never cease trying to be the best you can be. One is under your control the other isn't."

JOHN WOODEN[8]

While our job is not to judge if another person's definition of success is right or wrong, it will be obvious to us as we listen in conversation with them if they are being their own worst enemy. Comparisons can challenge us; but if they become that by which we define ourselves, we are in a never-ending race we will never win.

"Success is not final. Failure is not fatal. It is the courage to continue that counts."

WINSTON CHURCHILL[9]

DAY 10
CREATING PERFORMANCE

"I discovered the opponent in one's own head is more formidable than
the one the other side of the net."

Tim Gallwey[1]

I have had the opportunity on two occasions to sit under the teaching of
Tim Gallwey, whose work has often been credited as the foundation of
performance coaching. In the Endnotes, you will find several references
to his books, and other resources, that understand the work of Tim
Gallwey, which is based on a concept best known as the "Inner Game."

If you have attended several coaching workshops and never heard
of the Inner Game as it relates to coaching performance, please keep
looking for a workshop that does. If you are the steward of leadership
coaching programs within your company and the content does not include the impact of the Inner Game on coaching, consider making the
link and bringing it into your programs. If you consider yourself a master
coach but are not aware of the impact of the Inner Game on your coaching, consider continuing your coaching journey by exploring the Inner
Game concepts.

From a coach's perspective, what are keys to performance?

- People know what to do.
- People do what they already know.

- When people do not do what they already know, we look for what is keeping them from doing so.

If people do not know what to do, we direct them to the knowledge, training, and experiences so they learn what to do. From there, most people most of the time will perform by doing what they know to do. Coaching comes more into play when people know what to do but are not doing it. This is where Gallwey and his insights into performance can be a tremendous help.

> "Coaching is helping the person identify the interference and then replace it with a more helpful point of focus, resulting in performance."

In his early years as a tennis instructor, Gallwey noticed that if he stopped telling players what to do and instead demonstrated a move and asked them to mimic it, without verbal instruction, the player's performance improved. Too many specific instructions such as how to swing, where to stand, and when to move, swirling around in the brain, got in the way of what the players were perfectly capable of doing on their own. Gallwey called this swirling around in the brain "interference." Interference is anything that blocks us from acting on what we already know. Interference keeps us from performing in the moment at our best. Interference inhibits (gets in the way of) excellence in performance.

Gallwey simplifies it in a formula:
Performance = Potential − Interference

Potential is what the person can do. Performance is what the person does. If performance is not where it needs to be, interference is present and getting in the way. The sources of interference can be internal (things I think within my mind) or external (things outside that are beyond my control). Coaching is helping the person identify the interference and then replace it with a more helpful point of focus, resulting in performance.

Look over the list below. Identify the source of each interference which can inhibit excellence in performance. Which are internal interferences? Which are external interferences? There is no answer key. You are on your own!

__ Fear of failure
__ Policies and procedures
__ Trying too hard
__ Working to impress
__ Lack of technical, job knowledge
__ Ideas rejected
__ Lack of resources, tools
__ Seeking perfection
__ Compensation plan
__ Verbal instructions, telling

__ Lack of confidence (self-doubt)
__ Poor economy
__ Lack of opportunity
__ Anger and frustration
__ Lapse in concentration
__ Self-criticism
__ Unclear expectations
__ Excessive workload
__ Nervousness

InsideOut Development provides helpful insight from their "Study of Workplace Interference."[2] Below is a brief summary of their findings. In the Endnotes is additional information to locate the full study. You will also find other valuable coaching resources on their website.

- 60% of employees experience some degree of workplace interference every day that prevents them from accomplishing their goals.
- The average employee shows up to work only 68% charged up and ready to go.
- The sources of *internal* interferences identified most often include concern about money, worry about friends and family outside of work, worry what other people think, and concern if what they do really matters to the company.
- The sources of *external* interferences identified most often include disorganized workplace, distracting co-workers, difficult clients, and workplace politics.

- Helpful things managers can do to reduce workplace interference include:
  - o Treat employees fairly, equally.
  - o Provide opportunities for growth.
  - o Pitch in and help when things are hard.
  - o Coach employees to solve problems.
  - o Offer more training.
  - o Ask employees what they should do instead of telling them what to do.
  - o Take an interest in the personal lives of employees.

"Every job, every task is composed of two parts, an outside game and an inner game. Neither mastery nor satisfaction can be achieved without giving some attention to both games."

TIM GALLWEY[3]

DAY 11
IDENTIFY, FOCUS, REMOVE, IMPROVE

"The ability to focus the mind is the ability to not let it run away with you. It does not mean not to think—but to be the one who directs your own thinking."

TIM GALLWEY[1]

Let me summarize the Inner Game, as I best understand it, and reflect on how it fits into the coaching conversation.

1. If performance is not where it needs to be with a person, interference is present and is impacting performance. Assuming the person knows the expectations and understands them, start to identify the possible interferences.

2. The sources of interference can be internal (things I think within my mind) or external (things outside that are beyond my control).

3. Specific sources of interference at work might be:
   - Lack or doubt of ability to do the job (can be external or internal)
   - Things in the work environment (external)
   - The want to or desire to do the job (internal)

4. Step into each identified source and have a coaching conversation with the person to increase awareness of which source might be hindering performance the most.

   Example: If the source identified is concern of ability to do the job, below are questions that might be helpful in conversation with the person to help them think through the interference:

- What about the job are you comfortable doing?
- Where are you getting stuck?
- What would enable you to do that portion of the job? (Job knowledge, on the job training, mentoring, and practice are examples.)
- What are the next steps?
- What do you need from me?

These questions can be used with any of the interference sources that you identify.

5. Once the interferences are clear, sharpen focus on where the performance of the person needs to be, causing interferences to fade. Remember our **+** and fading circles picture?

"Sharpen Focus, Reduce Interference, Improve Performance!"

The **+** is symbolic of the performance goal, where performance needs to be. The fading circles are the sources of interference. Sometimes the person is focused on the goal but needs to sharpen the focus. Sometimes a person is not performing because they are focused on the source of interference and not the goal. In this situation, a change of focus is needed. For example, Bobby Jones, one of the early great golfers, would limit interference by playing against "Old Man Par" rather than focusing on his opponents. Competition with others was an interference to his performance.

"If I can beat the tar out of Old Man Par, I should do well on the leader board."

BOBBY JONES, AMATEUR GOLFER AND CO-FOUNDER OF THE MASTERS TOURNAMENT[2]

The coaching conversation is the vehicle for achieving sharpened focus or a change of focus. While understanding the product of coaching (FARM) and the concept of interference is helpful, don't let them become an interference to the coaching conversation. As you master curiosity in conversation, the specifics of FARM and the power of removing interferences naturally take care of themselves. As a coach, focus on curiosity in conversation is the key, not what you are trying to achieve (FARM). I might need to repeat that statement:

As a coach, focus on curiosity in conversation is the key, not what you are trying to achieve (FARM).

Real and lasting change in performance comes only as interferences are identified and replaced with focus on that which allows people to apply their best on the task before them. Not only are they now able to perform better, but they will also be more willing to take responsibility for the outcome as they have renewed ownership of their path forward.

Sharpen Focus, Reduce Interference, Improve Performance! Ensure that expectations are clear, and then hold the person accountable for the performance.

"Obstacles are those frightful things you see when you take your eyes off your goal."

HENRY FORD, FOUNDER OF FORD MOTOR COMPANY[3]

DAY 12

NICE IS NEVER ENOUGH

"Coaching is not merely a technique or set of tools to be pulled out and rigidly applied in a certain situation, but is a way of managing, a way of treating people, a way of thinking, a way of being."

JOHN WHITMORE[1]

I like nice people. I like nice leaders. Don't we all? The dictionary defines "nice" with words like pleasing, agreeable, pleasant, courtesy, politeness, respectable. I don't think any of those words or behaviors harm the world we live in, and often people with those traits bring sunshine into what otherwise may be a bad day. Is nice in leaders somehow a negative? No. But being nice in and of itself is rarely enough to make a credible leader-coach.

I think of two types of leaders when I think of Nice Is Never Enough. The first type fits the definition of "nice" in many ways. People enjoy their company and value the contributions they bring to the team, but they lack two traits or behaviors needed by a leader-coach: transparency and the desire to create connection with people. The lack of these two traits does not make them a lesser human being, and what they do well in their personal and professional lives most likely brings great value. Yet, when they are in a leadership role, people reporting to them often feel empty and cannot find clear answers to important questions:

- Who is this person I report to?
- What is important to them?
- Where can they help me grow?
- Do they know or care that I exist?
- Do they know how I bring value in my job?

When a leader clearly fits this type, a decision should be made very quickly to put them in a role where they can create value while no longer diminishing others with the feeling of not being valued. Outside the leadership role these "nice" people are likely to find more fulfilment in life, and all those who used to report to them can now find renewed hope and confidence going forward.

> **"Coaching tools and techniques do not make a coach."**

The second type of leader that comes to mind when I think of Nice Is Never Enough is the leader who most often makes great connection with others. They make time to invest in people, having a genuine interest in others, who they are and what they do. These leaders are often transparent to the degree that people find them authentic. For the most part, people enjoy working for them. What then is the problem? Their niceness can get in the way of their ability to provide clear expectations and feedback when expectations are not met.

Each of us needs feedback on how we can become better at what we do. We need leaders who will tell us the good, the bad, and the ugly of how we are perceived and will tell us, in a helpful way, when we are not meeting agreed upon expectations. Those expectations need to be stated clearly, and we do a disservice to others if we do not hold them accountable to the expectations. Those who are naturally direct and clear can learn to be "nice," if they are not already, but the journey of moving from "nice" to coaching with more directness and firmness is a tough journey, often too uncomfortable for those who love connection and harmony. While this second type of leader does not destroy the feeling of value, they can limit the growth, stretch, and challenge of those they are responsible for coaching. At some point, this leader may also need to be moved to a role where they can add greater value.

Coaching tools and techniques do not make a coach. A coach is who you are as a person and who you are with others. We are all blessed by nice people; but in leadership, Nice Is Never Enough.

"Only a life lived for others is a life worthwhile."

ALBERT EINSTEIN[2]

- Remember, in a leader-coach role, it is no longer about you. It is about the success and motivation of others.
- When you cannot see the good intent behind someone's actions, immediately contact them to better understand.

I have failed in this last one at times by being too quick to assume I knew the intentions behind someone's actions, normally assuming harmful motives or intent. The opposite, being generous in the interpretation of others' intentions, can make a significant difference in relationships with bosses, co-workers, and direct reports. We cannot read the intent of a person's heart, and good intentions may not always bring the desired impact, but generosity as a consistent practice can save a lot of harm and even conflict.

> **"***If you really want to see who people are, all you have to do is look.***"**

Maybe organizations, teams, and even families can find hope in such a simple, yet powerful, practice as showing kindness to one another!

> *"Life's most persistent and urgent question is, 'What are you doing for others?'"*

REV. DR. MARTIN LUTHER KING, JR.[3]

> *"Do all the good you can.*
> *By all the means you can.*
> *In all the ways you can.*
> *In all the places you can.*
> *At all the times you can.*
> *To all the people you can.*
> *As long as you ever can."*

JOHN WESLEY[4]

"There are three ways to ultimate success:
The 1st way is to be kind.
The 2nd way is to be kind.
The 3rd way is to be kind."

HENRY JAMES AND LAST WORDS BY MISTER ROGERS[5]

"You can't live a perfect day without doing something for someone
who will never be able to repay you."

JOHN WOODEN[6]

"He has told you what is good and what does the Lord require of you
but to do justice, and to love kindness, and to walk humbly with
your God?"

MICAH 6:8 (NRSV)

DAY 14

GO DOWN SO OTHERS CAN GO UP

"He must stoop in order to lift; he must almost disappear."

C.S. Lewis[1]

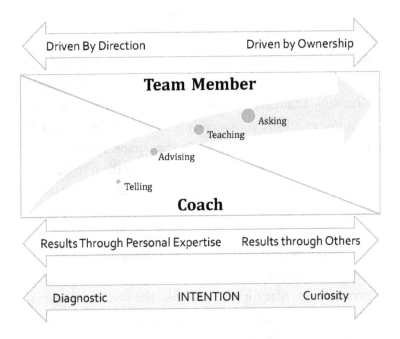

The model above, called the Coaching Continuum,[2] is an excellent picture of the coaching journey for both the coach and the person being coached. Also, we can use this model to do some soul searching regarding our fundamental intentions as a coach.

When facilitating a coaching workshop, I ask participants to study the model and discuss what they believe it is teaching. We are not able to have that dialogue here in this book, but let's dive into the learnings of this model in a coach-like manner. Below are some questions to ponder as you study the model:

1. What is the desired journey for the Team Member? (In this case, Team Member is the person being coached.)
2. What is the desired journey for the Coach?
3. What does the Intention Continuum at the bottom teach us about the importance of the Coach's intent?
4. How are Telling, Advising, Teaching, and Asking linked to coaching?

Below are my thoughts on the questions above. What would you add from your observations of the model?

1. As a coach, our goal is to move the person being coached toward greater learning, awareness, thinking, and responsibility of their situation, challenge, or growth. The way we coach can impact that journey. If we consistently lean toward Telling and Advising, then we create a dependency in the person to wait for direction. They get the message that we do not want them to think so much as we desire that they just do as they are told.
2. Results through personal expertise is the history of every leader, prior to their first leadership role. Now all that changes. Now it is all about getting results through others, not being the smartest person in the room, and moving away from a dependency on personal expertise for success. I like the saying, "Rather than get work done through people, start getting people done through work." It is all about helping others grow.
3. As we reflected on Day 3, diagnostic habits are engrained in us. Companies hire us to solve problems. Why else would they hire

us? Over time we develop a subject matter expertise that is recognized and rewarded for years. As a coach to others, staying with these diagnostic habits impacts our intentions and habits as we coach. We continue to examine the problem, seek the answers, and strive to fix things, which causes us to seek the awareness in the coaching conversation that belongs only to the person being coached. The journey of the coach is to move away from the diagnostic intent toward curiosity. If helpful, go back to Day 3 to review Curiosity in Conversation.

4. The journey from Telling to Advising to Teaching to Asking is the task of guided focus, awareness, responsibility, and mobility (FARM). As a coach moves from TELLING to ASKING, the person being coached moves toward more responsibility and ownership, while the coach moves away from solving issues using their own personal expertise, and stays with curiosity in the conversation. The person being coached becomes greater; the coach becomes lesser.

What else do you notice from the model?

What are your intentions as a coach?

When coaching, often ask yourself, "Who is discovering, learning and growing in awareness so they are able to make better choices?" You, as the coach, or the person being coached? Remember, it is always about the person being coached moving to another place.

> "*The person being coached becomes greater; the coach becomes lesser.*"

Coaches go down so others can go up.

DAY 15
SOMETIMES WE TELL, MOSTLY WE ASK

"The teachers who are very limited as conducting teachers are the ones who say, 'Do it exactly as I do, and you'll get the same result that I get.' That's a big mistake. If you force your students to toe the line in a mimicry sort of way, what you end up with is a lot of bad imitations of yourself."

KEITH LOCKHART, CONDUCTOR OF THE BOSTON POPS ORCHESTRA[1]

As we move from Telling to Advising to Teaching to Asking, take notice of the image below.

Sometimes We Tell, Mostly We Ask

Coaching Conversations

The full range of coaching conversations can fall under this umbrella. It is important for a coach to know when to tell and how to tell well with a coach's intent. It is important to know when not to tell. When you are telling or even giving advice, ask yourself, "Why?" Is it for the feeling of showing what you know and taking satisfaction in giving the advice, or is it in playing the role of the teacher or rescuer? Is it because you are uncomfortable with the silence that sits in the air as the other person thinks before responding, so you start to talk again so the conversation feels more comfortable to you? It is so easy to go to the role of rescuing others and sharing our expertise because it feels helpful; and we have been rewarded for doing so in the past because problems get solved and results come in more quickly.

Do you know when you move to telling or giving advice? How can you know? Tim Gallwey reminds us that *"Before we go about trying to change something, always increase awareness of the way it is now."*[2] On Day 4 we reflected on awareness and its impact on being in presence with others; but because of its powerful role in effective coaching, it will not harm us to reflect some more on the topic. Self-awareness is quite amazing, in fact, almost magical because it is self-fixing, assuming the desire to get better is present. To change a habit, we must first take notice and discover what we are currently doing. Do not try to fix anything, simply take notice. For example, if you want to know if or when you move to telling, take notice over several coaching conversations with others by taking two minutes after each conversation to exam what just happened in the conversation:

- What did you do?
- Who did most of the talking?
- Did the person leave committed to action?
- Did they discover how to get where they need to go because you told them or because you helped them create choices?
- Did the conversation cause them to think?
- Did they seem stuck at some point and you offered one or more solutions for them?

- Did you stay with curiosity, or did you move to diagnostic intent?

The questions you use can be different than the ones above, but the key is to take a few moments to make a few notes on what happened and then move on with your day. Do not over think the conversation throughout the day or focus on how good or bad you did. For a few weeks, keep the same routine after each coaching conversation; before you know it, you become more and more aware of what you do in a coaching conversation. Do not focus on fixing but on increasing awareness; and from awareness flows improvement. With the new awareness, you will be at a much better place to practice the balance of "Sometimes We Tell, Mostly We Ask." Whether you tell, advise, teach, or ask during the conversation, it will be with intention and with awareness that what you are doing is what the person most needs from you at that moment.

> *"We cannot change what we are not aware of, and once we are aware, we cannot help but change."*

SHERYL SANDBERG[3]

My mentors in coaching over the years have encouraged me to often ask myself during a coaching conversation, "What is the most useful, most valuable, most powerful thing I can do right now to move the person forward?"[4] If in the moment they need you to lean toward telling to move them forward, then go there. The telling or the giving of advice can be appropriate and helpful. But if the telling is in any way for your benefit or starts to transfer the ownership to you from the person being coached, then stop.

When a specific moment makes telling almost a given—urgent matters, crisis, high risk decisions—always come back later and have a follow-up conversation so you can hear the person talk about what they did, what worked, and what they would do differently next time. When you give advice, always cling to the advice very lightly. If you give advice to help create more options or choices for the person, be indifferent whether the advice is taken or not by the person. Ask "What else?" after giving advice to signal to the person that you offering advice does not stop their continued thinking through options and choices.

> **"***Do not focus on fixing but on increasing aware-ness; and from awareness flows improvement.***"**

Michael Bungay Stanier, author of two very practical coaching books, offers insight on the coaching conversation, *"Be curious a little longer, rush to advice-giving and action a little more slowly, and make coaching conversations a daily, informal act."*[5]

Sometimes We Tell, Mostly We Ask; think of it as the toggle switch of coaching. Use telling only as it is needed to move the person forward. Toggle back to asking to create pause and thinking. Continually grow more comfortable in asking for there lies the power of coaching and maybe our greatest challenge, as John Whitmore reminds us, *"It may be harder to give up instructing* (telling) *than it is to learn to coach."*[6]

> *"The strongest human instinct is to impart information, the second strongest is to resist it."*
>
> KENNETH GRAHAME[7]

DAY 16

WE SEE WHAT WE SEEK

Winnie-the-Pooh:	It's called, "Say What You See." You first, Piglet.
Piglet:	Panic, Worry, Catastrophe
Tigger:	Speed! Danger! Recklessness!
Eeyore:	Disgrace. Shame. Humiliation.[1]

We all look through lenses which determine what we see in life. While we can change our lens, we always look through a lens which influences our thoughts as well as our actions. You and I can look at the same thing and see things differently because of the lens through which we are looking. You have heard the saying, "Looking at the world through rose-colored glasses." Everything you look at looks that color. In the same way, things which we hold deeply flow from the lens through which we look at the world.

Sometimes the lens we look through for certain topics may not even be conscious to us or may not be something we can articulate. Sometimes it needs to be challenged. Also, life experiences impact our lens, assuming we permit ourselves to be challenged. In the broader scheme of life, this lens impacts our view of religion, science, and politics and reminds us that nothing in life is neutral or approached without bias. We are always biased and never neutral.

When related to coaching, the lens we look through can have a significant impact. One example are assessments and how we use them with individuals and with our teams. Assessments such as MBTI, Disc, Social Styles, StrengthsFinder (and the list can go on and on) can each be beneficial to individuals learning more about themselves and teams working more productively together. The challenge develops when a coach begins to look at people through the lens of a specific assessment.

When a coach begins to engage a person and view them as defined by an assessment, the potential for breakdown of connection and trust can accelerate. I recall a situation when it was necessary to transfer a team member from a supervisor who had provoked the relationship beyond repair because the team member's performance was always judged by the supervisor through the lens of how a specific assessment had defined the team member. The team member, in the mind of the supervisor, was simply an assessment. To avoid this trap that can come with the use of good assessments, keep in mind that you are never coaching an assessment. You are always coaching a person, and that person is seeking to be someone and do something. Assessments can help you know them better, understand why they do certain things, and give common terminology to enhance development discussions; but the assessment is never that person.

To use a simple analogy, with hopes not to insult anyone, envision people as a coat rack. The coats on the coat rack serve a purpose based on the weather outside, but they are not the coat rack. Likewise, people on your team come with assessments, but they are not the assessments. The assessments people have taken can be helpful to build connections and in development discussions, but it is the person we are always coaching and helping to grow. Also, if the person is not impressed with a certain assessment, it is best to avoid using that assessment in development discussions even if it is a corporate initiative to use the assessment. People must connect to the assessment and see it as a helpful mirror to look into if a discussion regarding it is to be of any value. As a leader, avoid the temptation of explaining away improvement opportunities for yourself by claiming, "That is just the way I am," and then using an assessment to justify poor leadership behaviors.

Is there a specific coaching lens, a bias, that can aid coaches in helping others to discover, learn, and grow? If we reflect back to Days 2, 10, and 11, we can create a powerful lens for coaching. A coaching lens that is bifocal, with one lens being "Talent" and the other lens being "Interference," can impact performance in a big way. Remember, we see what we seek. A coach who has a bias, a belief that talent is there to be found, will find talent and be consistent in using that talent to build strengths and excellence in others. A coach who is looking for the interference that is hindering performance will help identify the interferences and through coaching conversations help the team member remove them and move to better performance.

> **"**Consider the bifocal coaching lens of **talent** and **interference."**

A few years ago in a meeting with first level supervisors who had recently completed a coaching class, they were asked what coaching is and how it can help them in their roles as supervisors. One of the supervisors quickly spoke up and said, "My daily job is to look for interference that keeps my team members from doing the best they can at work that day. Find it and then remove it." You can imagine the smile that came to my face! It is so easy to leave a coaching class with good tools and having experienced good engagement in class but then back on the job still wonder, "What am I supposed to do different now?" That supervisor had discovered the simplicity of bringing coaching into his world to make a difference. He was looking at his team and the work environment through the lens of interference. As he looks for it, he is more likely to find it and help others remove it. Once the interference is removed, the talent present within the person has a better chance of surviving and performance is the result.

To be honest, that was the first time, after decades of facilitating coaching classes around the world, that I heard a coach articulate the simplicity of coaching in those words. I think that says a whole lot more about me than it says about the coaches who took my classes over the years. Maybe in time, I too am discovering the essence of coaching!

What lens are you looking through today? You are looking through a lens, a bias of your belief about people and work. Consider the bifocal coaching lens of *talent* and *interference*. Not only will you find talent and interference, but you will also create performance.

"Imagine what is possible with access to all the intelligence that sits in your organization."

LIZ WISEMAN, CEO OF THE WISEMAN GROUP, RESEARCHER, AUTHOR[2]

DAY 17

SOME THINGS TAKE NO TALENT

"You can't always be the strongest or most talented or most gifted person in the room, but you can be the most competitive."

PAT SUMMITT[1]

We have reflected on the importance of looking for talent in others. It is half of our bifocal coaching lens. Using talents in our work is a great source of personal job satisfaction. In the business world, people working with their talents is a strong source of competitive advantage.

However, whatever happened to basic job expectations and asking people just to do their jobs? Well, there is still power in asking people to show up and do a number of things that take no talent: TNT - Take No Talent!

I will start a list of TNTs, and you add to it:

- Being on time
- Being prepared
- Making effort
- Displaying energy
- Knowing expectations
- Following procedures
- Communicating
- Keeping focus

- Showing enthusiasm
- Caring
- Trusting
- Believing
- Being coachable

And what about just Thinking!

Coaches desire to build commitment in others, for with commitment comes ownership. Along the journey to commitment, however, it is important to expect compliance. Seek commitment, but expect compliance until the commitment matures.

Do not hesitate to have expectations of people beyond just showing up at work, and always hold them accountable to give the things that Take No Talent.

> "*Seek commitment, but expect compliance until the commitment matures.*"

> "*Never make excuses. Your friends don't need them, and your foes won't believe you.*"

> JOHN WOODEN[2]

DAY 18

WE DON'T MAKE ANY NOISE

"It's very important to realize that you don't make any noise. That you are a silent partner in this thing. If the conductor forgets that and starts to think that he is *creating the music*, you begin, I think, to undervalue the contribution of the very talented musicians who are under your command."

KEITH LOCKHART, CONDUCTOR OF THE BOSTON POPS ORCHESTRA[1]

Coaching is a part of leadership, assuming the leader desires it to be. It is not automatic, but a choice. The best leaders do a lot of things well, but when leaders purposefully add coaching as a key component of their leadership brand, they experience making a difference in a way unique from other aspects of leadership. Multiple books and articles advise leaders to initially step into their leadership roles with the mindset of, "Once in a leadership role, it is forever no longer about ME." Regardless of the management level of the leader, it is clear which leaders operate from this mindset and which ones never took this council to heart but are still focused on themselves. The latter desire to create the music but are often their own worst enemy (internal interference) in doing so. Choosing coaching as part of your leadership brand moves your focus away from yourself and makes the future about the growth and success of others.

Over the years, I was challenged by senior leaders to value a broader range of leadership styles within the organization. As you might imagine after reading several days of my reflections, it was easy for me to look for and value leaders who build connection with others and greatly value their role as a coach. I had a bias that needed to be challenged. Recently, I came across a quote that reminded me of that challenge from an observer of two British Prime Ministers as the start of World War II was becoming inevitable; *"If I had to spend my whole life with a man, I would choose Neville Chamberlain, but I think I would sooner have Mr. Churchill if there was a storm and I was shipwrecked."*[2]

Often teams, or the broader organization as a whole, face unique challenges that call for different leadership styles, approaches, and unique personalities, each with their unique strengths. I have learned to value a broader range of different leadership styles and personalities and not to judge certain leaders unfairly from afar. The impact that any leadership style has on people and the organization can be positive when kept in balance. Coach John Wooden wrote, *"Next to love, balance is the most important thing."*[3] In an earlier daily reflection, we reflected on Nice Is Never Enough in leadership. If the "Nice" leader shows indifference toward people, or a focus on accommodating others, people are not challenged to make great music, and the desired results are not achieved. But "Nice" with a balance for achieving results can be productive for everyone.

On the other extreme is the "results #1 priority" style and the damage it, without balance, can bring to individuals and the team. There are clearly times when a more command and control, all-hands-on-deck, result-focused style is needed for the challenge ahead. A leader who naturally leans toward this style typically has strengths for the challenge before them; but without balance this style of leader can become very comfortable in consuming most of the space that could be given to others' thoughts and ideas. While they normally are the smartest person in the room, if their self-awareness is low, the pain comes in watching them display how smart they are again and again and again–up, across, and down the organization!

Over time this "results #1 priority" style will often develop a proven record for reaching targeted results. (Remember, where you focus is where you go.) Leaders with this style will get things done. Often that comes from their ability to know where the "bottom of the slope" (the goal) is, and the clarity in which they communicate expectations. Then, they consistently hold everyone accountable to the expectations, an extremely powerful and valued trait in a leader. Yet without a pause to look into the mirror and balance the bigger picture of themselves as leaders, they will begin to believe they can create the music alone. As Keith Lockhart, in his role as a conductor, reminds us in the quote above, *"To undervalue the contribution of the very talented musicians who are under your command"* is the greatest source of failure for a leader who otherwise has the ability to create beautiful music with others when focused on their growth and success.

> "*In leadership we lose the right to put ourselves first.*"

If the leader gets lost in the belief that they can create the music alone, their focus, energy, connection, and time turn toward the level of management that is looking for the results, and turn from those who need to follow their leadership (the "musicians"). The result is that the leader is often perceived as inconsistent and unpredictable. Then when results are not forthcoming, punishment, even if only verbal, is often the fall-back technique to motivate people to get the desired results since there is no foundation of connection to motivate in other ways. With the pressure of missing targeted results often come a range of negative interactions with followers and support organizations. These interactions can include dismissive comments to others' ideas, personal feedback that harms rather than helps, and sometimes, as emotions escalate, the degrading of people in a team or public setting.

We all have bad days when we get frustrated, emotional, and aggravated over things not moving the way we know they could if we took control. Leadership talent which delivers on results is to be valued. Yet, without balance, the leader will eventually crush the confidence of others and increase disengagement as people step away from the work to protect themselves. The talent for achieving results with a balance for creating great music through the engagement of the many can be productive for everyone.

Simon Sinek says it wisely and briefly: *"The cost of leadership is self-interest."*[4] In leadership we lose the right to put ourselves first. That does not mean ambition is evil or all about self. It is not poor leadership for a leader to be confident and assertive, with specific career goals. To me, the difference is how much of oneself consumes one's focus, time, and attention. C.S. Lewis brings a perspective to ponder: *"True humility is not thinking less of yourself; it is thinking of yourself less."*[5]

Without balance and the proper sense of humility, even the best of leaders, as they start to enjoy the success of leadership, may get caught up in being right and being told by others that they are the smartest person on the planet. What can happen next? They stop listening to the input of others and stop seeking the brilliance in others to fill their knowledge and experience gaps. Learning stops, and *"when you are through learning, you are through."*[6] Humility enables us to step behind the curtain and let others succeed and shine. With humility comes focus on the leadership opportunity before us rather than on the need to prove ourselves to others. We are freed from the constant need of self-promotion and find it more satisfying to pour ourselves out for someone else's gain. The sooner a new leader gains this perspective, the more likely they create a balanced leadership style that draws followers to their team, enhancing opportunities for their advancement. (Remember, Nice Is Never Enough, But Kindness Is Everything!)

Lord Beaverbrook, minister of aircraft production under Winston Churchill, described Churchill's leadership in this way: *"He could get very emotional, but after bitterly criticizing you had a habit of touching you, of putting his hand on your hand as if to say that his real feelings for you were not changed. A wonderful display of humanity."*[7]

You may have noticed that I dedicated only a few sentences on balance and the Nice Is Never Enough style, while dedicating several paragraphs on balance and the "results #1 priority" style. Do I still have a negative bias toward the result-focused leader? No! I now better understand its power to impact for the positive and how quickly that impact can turn to the negative. The potential of a result–focused leader to elevate and challenge individuals, the team, and an entire organization is very high. People can experience productive challenge and growth like never before. Yet, the potential to diminish engagement and increase burnout is as equally high. The talent and strength needed to drive toward results can, when overplayed, become a powerful, visible weakness. Leaders with a bent toward the result-focused style need to spend more time reflecting on balance and remembering that being a leader who can see the path to great results does not need to stand against or in place of the many talents of the people on the team and within the organization.

Liz Wiseman summarizes the choice of leadership style very well: "*Some leaders literally shut down brain power in the people around them. Yet, other leaders seem to amplify the intelligence of the people around them.*"[8] Which choice have you made?

No leader has a perfect leadership style. That is what makes leadership, as well as coaching, so fascinating, for it is imperfect people trying to have an impact in an imperfect world engaging with imperfect people. No single formula guarantees success. Not every leader fits the need of every leadership role. The leadership style of the leader needs to match the "need of the day." Failing to consider the proper match can lead to leadership failure, when in reality the leader may be very effective leading another team or organization facing different demands and challenges.

It takes a unique person to bring the courage needed to create leadership balance. (*"Courage is what it takes to stand up and speak. Courage is also what it takes to sit down and listen."*) Do the people connected to your leadership feel a genuine sense of value, contribution, and accomplishment? *"When the best leader's work is done, the people say, we did it ourselves!"*[10]

"I would rather believe in people and be disappointed some of the time than never believe and be disappointed all the time."

JOHN WOODEN[11]

"My job is not to be easy on people. My job is to make them better."

STEVE JOBS[12]

"It's what you learn after you know it all that counts."

JOHN WOODEN[13]

"Talent is God-given. Be humble. Fame is man-given. Be grateful. Conceit is self-given. Be careful."

JOHN WOODEN[14]

DAY 19
EVERYONE HAS A STORY

"Each of our stories are inside a very great story."

J.R.R. TOLKIEN[1]

If my world travels have taught me anything, it has been that everyone has a story and their story is nothing like my own story. You are a story, and so am I. We are not snapshots but movies. We are all part of a larger story. No one's story is normal for it is uniquely their story and composed of moments of great joy, as well as disappointments, along the way. While we want to avoid imposing our story on others, we should feel free to bring it to our conversations. As we live in each other's stories, we also live in our own story.

What is my story?

I was raised on the coast of Virginia, racing sailboats most of my young life. I married my high school sweetheart at the young age of 19 years old. We raised five children together and today find tremendous blessings in seeing them live their lives. I have jumped out of helicopters to fight forest fires. I took my first international trip in 1984 to Uganda, East Africa, on a mission trip to help restore the health of local schools. In 1995 I renewed my passport and began 25 years of corporate global travel facilitating leadership classes and coaching in over 20 countries. My company of over three decades gave me the best job ever for me. I am blessed today with a growing family who loves me and friends across the globe. My lasting hope rests on the righteousness of Jesus Christ, not on my own merit.

Now is that really my story? Of course, not completely, but it tells you things that are important to me. Do you know something similar about each person you coach? Can you write a short paragraph about each person you lead? A person's story can be discovered only by caring enough to know their story. The best coaches continually nurture interest in human stories. While we are not always able to know, or understand, what each person is feeling within their story, we can be interested. It is not the job of those we lead necessarily to be *interesting* people, but it is our daily commitment to be *"interested"* in each of them. We find their story in conversation.

While each person will decide how much they want you to know of their story, I have seen coaches benefit from knowing the answers to questions similar to the ones below. Most often people are comfortable in answering them. Consider seeking the answers to these questions with the people you coach:

- What excites you most about your work right now?
- What aspect of your work is most frustrating to you?
- When are you at your best?
- What keeps you from giving your best?
- On your very best day at work when you go home and think, "I have the best job on the planet," what did you do that day?
- What have you learned about yourself in your current role?
- How do you explain what you do at work to your family and friends?
- What do you need from me that would make it easier for you to give your best?[2]

"It is not the job of those we lead necessarily to be interesting people, but it is our daily commitment to be "interested" in each of them. **"**

When it comes to yourself, can you answer the additional questions below within your story? Are you comfortable sharing your answers as it helps to build connection with others? The questions below are deeper than the questions above and may not always be suited for you to share at work, but the process of exploring the answers will increase your ability to be more transparent with others. People like working for real people, and the more they know your story the more they know that you have experienced success, joy, failure, and disappointment, just as they have.

- Who has touched your life?
- What is the YES that brings you delight?
- What are you against?
- What is your NO?
- What brings you to tears?
- Where do you laugh?
- What burdens you?

"Perhaps the greatest risk that any of us will take: to be seen as we really are."

Cinderella, 2015 Motion Picture[3]

One of my favorite stories about stories comes from Walt Bettinger, President & CEO at Charles Schwab, sharing a life lesson from his college business strategy final exam.

After hours of preparation for the exam, reviewing formulas and case studies, on the day of the exam, the professor handed out just one sheet of blank paper. The professor explained, "I have taught you everything I can teach you about business in the last 10 weeks, but, the most important question is this: What is the name of the lady who cleans this building?"

Bettinger was stunned and it was the only test he ever failed. In an interview with the NY Times, Bettinger said, "I got the final grade I deserved. Her name was Dottie, and I didn't know Dottie. I had seen her, but I had never taken the time to ask her name. I have tried to know every Dottie I have worked with ever since. You should never lose sight of people who do the real work."[4]

Forever burned in his mind and heart was the importance of respect for each person, regardless of their job level within the organization. Every person is made on purpose and with a purpose, created with inherent dignity, worth, and value, none of which is earned by a certain status obtained in life.

"It isn't what you do, but how you do it."

COACH JOHN WOODEN[5]

Be in awe of others' stories, live in gratitude of your own story, even the parts that trouble you. For it is the highs and lows in life, the good and bad, the joys and disappointments that make our stories unique and ultimately make us who we are.

"Tell me your story, I'll tell you mine.
Sing me your song, I'll follow line by line.
Draw me near, let me hear the things you've treasured.
Patient as fallen snow, standing inside the questions.
Only guessing by what truths our souls are measured.
Each of us rising from worlds unknown.

Within your trials, I see my own.
Still there are journeys that are yours alone.
You were born for the storm you have to weather.
True as a winter wind, you faced the moment bravely.
You and I we are on our own, and yet together.
Walking a path, we can't define.

Tell me your story, I'll tell you mine.
Sing me your song, I'll follow line by line.
Let the night fall with the lightness of a feather.
Trusting the coming dawn.
We cannot hold the morning.
You and I we are on our own, and yet together.
For in the end we are all flying home."

"FLYING HOME" (THEME SONG FROM THE MOVIE *SULLY*)[6]

DAY 20
HONOR EVERYONE

"The viable community; one that embraces healthy children, strong families, good schools, decent housing, and work that dignifies, all in the cohesive, inclusive society that cares about all its people, is a dream that lies before us. It is clear, that the dream will remain a dream until we move beyond the barriers we have built, consciously or unconsciously, around race, gender, equal access, and the composition of the workforce."

FRANCES HESSELBEIN[1]
FORMER CEO OF THE GIRL SCOUTS OF THE USA, PRESIDENTIAL
MEDAL OF FREEDOM

Diversity and inclusion initiatives and focus can be found in most every organization today. For too long, specific groups of people have been left on the sidelines, not able to bring their uniquely-designed talents and potential to the marketplace, or to society as a whole. I have chosen a day to reflect on this topic because of how helpful coaching can be in creating the right environment for successful diversity and inclusion.

Coaching, at its best, is stepping away from self-interest and stepping toward the interests of others. It entails knowledge of a person, interest in them as unique individuals with specific stories, capabilities, talents, and desires. Coaching that encircles a team or work environment can be the best setting for inclusiveness, a community where differences bind people together while also maintaining and cherishing what makes us different. We are a people created with the desire to seek connection with others, to know and be known. From what I have experienced, this holds true in cultures across the globe. Coaching eliminates separation and isolation, giving us the feeling of belonging.

Bates Communication has done excellent research on what contributes to high-performing teams and has created the Leadership Team Performance Index (LTPI) to help teams examine how their teams are performing. One of the fifteen facets that Bates found contributed to high-performing teams is Belonging, defined as "valuing and respecting differences, fostering an environment where all experience a fullness of membership and affiliation."[2]

Belonging is a feeling, the underlying target and foundation, of successful diversity and inclusion efforts. Diversity is the way we are different. Inclusion is the way we value each other and thrive in working with each other. Belonging is acceptance, a spirit of kinship–I matter, and I can be my unique self. Diversity and inclusion are divergent efforts intended to highlight and focus on the success of specific groups of people disadvantaged in the past. Belonging is convergent, bringing everyone back together with each person comfortable to bring the best version of themselves. Everyone is accepted for who they are and expected to be who they are in all settings. With belonging comes a voice for each person to be heard, ownership by each person, and accountability among persons. The team and work environment become a richer place for people to live and work.

When you are leading and coaching teams, make belonging a topic often discussed. Discuss what the team does well that helps others feel belonging, what is currently in the way of creating a stronger feeling of belonging, and what the team could do more of to create belonging.

"*Dignity, worth, and value are inherent in each person, and coaches should never allow one person to take them from another.*"

I do not have a single daily reflection dedicated to coaching teams but chose instead to reflect on teams here, on Day 20, within the topic of diversity and inclusion. I believe diversity and inclusion are the heart of effective teams. If the leader-coach does not expect and inspect teams to value not only the best work that people bring to the team but also the individual who is bringing that best work, teams will not be effective in seeking different points of view and engaging each other in productive conversations. John Wooden wrote, "*The main ingredient of stardom is the rest of the team.*"[3] There must be the expectation that the one team member is as important as the many; and within the success of the many, the one will find a special place to be themselves and add value.

Anne Frank, famous for the diary she wrote while in hiding during World War II, wrote, "*We all live with the objective of being happy. Our lives are different, but yet the same.*"[4] My travels across many cultures taught me the uniqueness of each culture, but I also saw the truth in Anne Frank's quote. We all start at different places and we should seek to understand those differences; yet across the world we are so much alike in the things that make us laugh and cry, and in how we love those most important to us. An over focus on our differences can lead to unproductive divisions between us.

Building teamwork is not an easy path. Pat Summitt wrote, "*Teamwork is a lot like being part of a family. It comes with obligations, entanglements, headaches, and quarrels. But the rewards are worth the cost.*"[5] It takes courage for each of us, when working with others to "*Consider the rights of others before your own feelings and the feelings of others before your own rights.*"[6]

People working and living together will always differ in certain ways, in belief systems and values; and those differences should be listened to and honored. If they are not, there is the risk that diversity and inclusion programs, over time, will only create a different group of people assigned to the sidelines. Dignity, worth, and value are inherent in each person, and coaches should never allow one person to take them from another. Honor everyone.

"Honor everyone."

I PETER 2:17 (ESV)

"We don't have to be superstars or win championships.... All we have to do is learn to rise to every occasion, give our best effort, and make those around us better as we do it."

JOHN WOODEN[7]

"If you want to go fast, go alone. If you want to go far, go together."

AFRICAN PROVERB[8]

"I was asked if I had any advice when it came to coaching women. I remember leveling him with a death ray stare, and then relaxing and curling up the corner of my mouth and saying, 'Don't worry about coaching women. Just go home and coach basketball.'"

PAT SUMMITT[9]

"Finally, all of you, have unity of mind, sympathy, love, a tender heart, and a humble mind. Do not repay evil for evil or reviling for reviling, but on the contrary, bless, for to this you were called, that you may obtain a blessing."

I PETER 3:8-9 (ESV)

DAY 21

IT'S ABOUT JOURNEYS

"The years teach much which the days never knew."

RALPH WALDO EMERSON, 19TH CENTURY WRITER, PHILOSOPHER[1]

This day of reflection will reflect a clear time stamp. I am writing during the 2020 global COVID-19 pandemic. In just a matter of days, the world went from life as normal to social distancing, quarantines, school closings, border closings, economic shutdown, shelter-at-home orders, lack of protective equipment to serve those dying from the virus, and many more never-seen-before responses across the globe to fight an enemy that we cannot even see–a virus. Life as we know has been put on hold, with no real clarity of when it will return or what "normal" will look like in the future.

The journey through COVID-19, like other significant events in past generations, will forever impact the perspective of those who live through it. Perspective comes from lived journeys, experiences. Only as one lives does one gain a perspective, a point of view. Perspective is certainly linked to the lens we reflected on in an earlier daily reflection, the lens through which we interpret our world, but the two are slightly different. My lens, the bias that I bring to my world, determines how I interpret, for example, the COVID-19 pandemic. The journey of living through the pandemic will shape my perspective. Hopefully, if I am going to maintain my effectiveness as a coach or, more important, as a person in community with others, the new perspectives gained from living through the pandemic, as well as other life journeys, will impact my lens, challenge my bias, help me interpret my world differently, and, hopefully, make me a wiser and better person.

Each of us also gains perspective as we walk with others through their life journeys. They can equally impact how we coach, lead, and live. During my corporate career, one of my peers lost both of his sons within a three-year period. Another peer lost her son in a single moment while he was attending kindergarten during the day. One of my coaching mentors lost his daughter in a car accident. Within the past year, a lifelong friend lost his son, who was in the prime of his life, to a battle with cancer, leaving his wife and young daughter behind. These losses give us perspective over the years of what really matters in life and at work.

J.R.R. Tolkien, author of *The Hobbit* and *The Lord of the Rings* books, when asked in the early days of writing what his story was about, said, *"It's about journeys. Adventures. Magic, of course. Treasure. And love. It's about all kinds of things really. It's hard to say. I suppose it's about quests, to a certain extent. The journeys we take to prove ourselves. About courage. Fellowship. It's about fellowship. Friendship. Little people just like you."*[2]

The journey of life sometimes brings harsh experiences, when it feels like the sun will never shine again, or that the road we are traveling seems to be going nowhere. Yet, other times are full of hope and possibility, reminding us of a beautiful blue-sky day with flowers blooming. These journeys teach us day-to-day lessons which over time develop our perspective. Perspective is not gained in a day but over time. Our journeys in life take us many places and give us many experiences; some we may wish we had lived without. Yet, they are a part of our story, the person we have become, and we are all part of a greater story. New journeys lie before us and for those we coach. These journeys are not just experiences to write down in a journal, but they shape who we are and how we coach. Journeys, if we listen to them, give us perspectives that help us become less harsh, more patient and compassionate in our response toward others. When we see others struggle, we recall when we have struggled. When we see others fail, we recall the times we have failed. My connecting with you often requires me to connect with something within myself that knows what you are experiencing.

My best coaches helped me to consider what was important along the way, to keep the big picture of family, friends, faith, and career success in perspective and with proper order of priority. They did not tell me what my priorities should be, but they certainly role-modeled their own set of values and priorities and honored the priorities I chose as important to me. They were open and transparent enough to be known, making me comfortable to be more open and transparent than I would have been otherwise.

> *"I want to run for eternal glory and track is great, but it's not what life is all about."*
>
> ALLYSON FELIX, ONLY FEMALE TRACK AND FIELD ATHLETE TO EVER WIN SIX OLYMPIC GOLD MEDALS[3]

As you coach others, the perspective you bring to coaching conversations can be a significant blessing in helping others make choices during the work challenges of the current week or life journeys they face along the way. Bringing perspective to a conversation is not telling, not fixing, not thinking for another person, nor is it taking ownership from the person. Bringing perspective to the coaching conversation is not standing in judgment of what you hear but is honoring another's perspective as their truth, which serves more as a mirror that helps the person to pause, reflect, learn, and think before they act.

> *"Journeys, if we listen to them, give us perspectives that help us become less harsh, more patient and compassionate in our response toward others."*

Coaches help others achieve many things: increase their self-aware-ness, learn and grow, achieve opportunities they desire, conquer flaws in their leadership behaviors, take a team to the mountaintop of success, and much more. Yet, at times there is the opportunity to bring some-thing even bigger to our coaching conversations, such as helping others avoid lives of "almost and could have." I have mixed emotions about that phrase you just read; "almost and could have." It sounds so insightful when I read it one day, and then the next day it sounds like a trigger for years of guilt and regret! Yet, as a coach, bringing perspective to coaching conversations can help people move through situations where they are stuck, situations in which over time they would regret staying stuck.

For context I want to restate a quote from Sheryl Sandberg that I used in a previous daily reflection: *"If I had to embrace a definition of suc-cess, it would be that success is making the best choices we can...and accept-ing them."* This is great reminder that life's journeys bring us moments of choice when we make decisions, and then move. We cannot live a life of second guessing all our choices, for what a painful life that would be. Yet, when I think of the phrase, "almost and could have," in the context of a coaching others, I think of two emotions that have caused people to get stuck, interfering with success in the workplace and laying a foundation for professional and personal regret: anger and bitterness, especially anger that grows into bitterness.

Assuming these two emotions are not too deeply embedded (remem-ber, coaching is not counseling), bringing perspective to the coaching conversation has the potential to help a person move through anger and not get stuck in bitterness. A broad range of emotions can be healthy; but when a negative emotion like bitterness finds root in the heart, a person can become defined by that negative emotion, impacting the effective-ness of their professional and personal life. Coaches help people move, and coaches need to help people keep moving so they avoid getting stuck.

Anger does not last forever and can often be a fruitful way to express concern about a situation to others and then let it go; but bitterness can consume us, creating tangled webs in our hearts and minds. Bringing perspective in these emotional situations may be as simple as reminding a person to ponder gratitude and forgiveness. (Remember, we see what we seek.) When we seek gratitude, we find more reasons to be grateful. When we seek forgiveness, and we find more reasons to let go of bitterness.

"Forgiveness. That was so easy. Why didn't I do it ages ago?"

C.S. Lewis[5]

Coaches can bring perspective to situations that help others face an immediate challenge, project, or even crisis while keeping the big picture or journey of life in mind. In bringing such perspective, accompanied calmness and insight can help others move to better decisions and practical next steps. Listen to what people say to you. Without any intent to solve the issue before them, do not be afraid to help them see their own experience in a slightly different light by bringing your perspective to the conversation.

"A coach's primary function is not to make better players but to make better people. The game of life takes precedence over any athletic endeavor. The most important lessons must last a lifetime, not just a season."

John Wooden[6]

"The grand essentials of life are something to do, something to love, something to hope for."

Joseph Addison, English Essayist, Poet, Playwright, and Politican[7]

DAY 22

FEEDBACK IS NOT COACHING

"Trying to improve or fix others is a futile effort that usually ends up annoying them and frustrating you."

AUTHOR UNKNOWN, CITED BY GINA SOLEIL, EXECUTIVE COACH & AUTHOR[1]

Feedback is not coaching, and coaching is not feedback! Yet, they are best when connected to one another. In the context of moving others to a better place, feedback is a powerful tool for coaches to help others thrive and excel. Picture a performance cycle with me, a cycle of conversations to help others increase performance:

1. Establish clear expectations and goals.
2. Observe performance, paying attention to how the person is doing.
3. Have the feedback conversation.

When we notice performance *consistent* with expectations or moving toward the goal, we have a positive feedback conversation with hopes that the person's behavior that we have noticed will continue. When we notice performance that is *not consistent* with expectations or not moving toward the goal, we have a negative feedback conversation with hopes that the person's behavior will move in another direction. The coaching world has several words to call feedback, words which nobody likes to hear: negative, constructive, corrective, improvement, for example. In the context of this book, feedback to change behavior is called *improvement* feedback, although I am convinced that people receiving the feedback will always call it *negative*!

> "*Coaches shape behavior and encourage along the journey to the destination.*"

4. Coach.

Marcus Buckingham and Ashley Goodall provide insight on feedback: "*Tell me where I stand, but in the context of helping me to get better.*"[2] Feedback is information about the past, given in the present, with the goal of influencing the future. Feedback, however, does not shape the future; the coaching conversation does.

C.S. Lewis wrote, "*I sometimes pray not for self-knowledge in general but for just so much self-knowledge at the moment as I can bear and use at the moment. Total self-knowledge is not necessarily the goal but rather knowledge to know what to do next.*"[3]

The goal of feedback is not to change ten things about what a person is doing. If the list of improvement opportunities is that long, that person may be in the wrong job. Consider a new fit for the person before terrorizing them with unfruitful feedback. Feedback is not helpful if aimed at trying to change who the person is as an individual. Also, if you generally do not like the person, please do not use feedback as a weapon. Feedback is most helpful when very specific, providing new awareness, and then followed by a coaching conversation to help the person determine what next step would bring the best improvement.

Last, do everyone a favor by not starting off your coaching conversations with the statement, "I am going to coach you," or "I need to coach you." Doing so will train the people you coach to dread what is coming next! Remember, you are a coach, and if you are in a leadership role, you are a leader-coach. Avoid looking at yourself or sending the perception to others that you toggle between the leader role and the coach role. The coaching conversation is not a "hat" that you put on before you start the conversation. The two roles are never divided for they are who you are, a person, undivided in the role of a leader-coach. So, when it is time to coach, simply step in and start a conversation to help the person determine their next best step in response to the feedback they have received.

5. Hold the person accountable for performance.

Reinforce progress toward the expectation or goal. Avoid being the coach who is satisfied only when people arrive at the destination. Coaches shape behavior and encourage along the journey to the destination.

Buckingham and Goodall give us some additional insight on why conversations are so important:

- When *others* do not behave as expected, we create stories about "who they are," that something is wrong with them, and we begin to plan how to fix them with candid feedback.

- When *we* ourselves fall short of expectations, however, we create stories and emphasize "what is happening to us." Example: When another person misses a deadline, we contribute their lateness to a personal habit, for example, of always being so disorganized. But when I miss a deadline, it is because of something outside of me that caused me to miss the deadline, for example, the week was full of sick children and car repairs. We are much more generous to ourselves than to others when interpreting behaviors.

- Only in conversation can we learn the context and story that helps us give the most effective positive and improvement feedback to others and then help them move to improvement.[4]

Always follow feedback, especially improvement feedback, with a coaching conversation.

"Feedback can be, in the wrong atmosphere, a code for a performance problem. In other words, you're calling it coaching, but it is really criticism. And good coaches don't coach that way."

HERB KELLEHER, CO-FOUNDER OF SOUTHWEST AIRLINES AND LATER CEO[5]

DAY 23
IT'S ABOUT ATTENTION, MORE THAN FEEDBACK

"People do not like feedback. What they want is attention."

Marcus Buckingham[1]

Attention itself can increase performance, and attention to what people do best increases performance even more.

Today's reflection focuses on one of my most significant learnings over the past few years. Like many things we learn, simply paying attention seems so simple and makes me wonder why I did not articulate it before others did! As with all the daily reflections in this book, you must decide how it will impact the way you coach others.

Some learnings in this arena are from many decades ago, yet no less helpful. The Hawthorne workplace study from Chicago in the 1930s taught us that whether positive or negative conditions were changed in the workplace, worker productivity went up. When the study was completed, any increase in the productivity went back down. Learning: what drove productivity was not the changes made in the workplace but that management was paying attention to the workers during the study.[2] Is this learning from 90 years ago no longer true in our high-paced, agile work environments?

We may find some help in answering that question by looking at a more recent Gallup study linking the impact of what a leader pays attention to with the level of engagement by team members:[3]

Study results:

- If you want to create disengagement among people, simply ignore them. Pay no attention to individuals, and do not acknowledge the value of their work. Disengagement can be as simple as passive resistance in the workplace or can be active behavior that works against improvement and growth.

  Result: Less than 1% of your team will be engaged.

"*Talk to me!*"

- If you at least talk to team members, even if your attention is primarily focused on correcting or fixing them, engagement of the team moves to almost 50%. Talking to team members is a form of paying attention and acknowledges that they exist. The resulting engagement may be closer to the compliance level, with a significant amount of discretionary effort not given, but team engagement does move from less than 1% to almost 50%.

  Lesson #1: Talk to me!

  Pat Summitt, on the importance of talking to your people, wrote, "*In the absence of feedback, people will fill in the blanks with a negative. They will assume you don't care about them or don't like them.*"[4]

- If you can move your attention to a dominant focus on strengths and recognition, active dis-engagement practically disappears from within the team. Team members will more freely give their discretionary effort and bring more of their best to work.

  Lesson #2: People crave attention and more so crave attention on what they do best.

To continue learning about the power of attention, search on YouTube for "Marcus Buckingham on Coaching vs. Feedback," 2:31 minutes in length. After listening, ponder the following questions:

- Are people so much desiring feedback, or do they crave attention more?
- What role does non-judgmental observation play in the performance cycle?
- Do you take the time to talk with the people you coach?
- Do you know what the people you coach do best?

In recent years, it has been a very helpful learning for me in my role as a coach that attention itself can increase performance, and attention to what people do best increases performance even more. I hope you will ponder how it might influence the way you lead and coach others.

DAY 24

YES, THAT! STOP THAT!

"Don't spend time beating on a wall, hoping to transform it into a door."

CoCo CHANEL, FRENCH FASHION DESIGNER AND FOUNDER OF
THE CHANEL BRAND[1]

To recognize, in its original form, means "to know again."[2] Recognition is a conversation in which we reinforce a person's effort and get to know the person better. We value individuals because they have inherent dignity, worth, and value. That comes free just for being a person. We reinforce individuals when we see excellence in their efforts and progress. In the context of performance, our recognizing the individual builds self-esteem, while recognizing individual *effort* builds self-respect. Those standing on a foundation of *self-esteem* shy away from mistakes, failure, and learning, for they harm the view they have of themselves. Those standing on a foundation of *self-respect*, on the other hand, have learned they can make something happen, so they are more likely to regard mistakes and improvement feedback as useful information to move forward and improve.

When giving recognition, share the specific excellence that you see others do. Then let the person talk about it so you learn more about who they are when at their best. Recognition is a conversation, not a quick and done "thank you." (Remember from Day 23, Talk to me!)

In 2005 I had the opportunity to be certified by Gallup as a StrengthsFinder Coach. Strengths-based coaching teaches that you get to strengths by starting with talent. Related to the coaching lens from Day 16, coaches are always looking for talent, for it is the foundation from which strengths are developed and excellence is achieved. Excellence is cultivated by building on the best of each person. When we discover a weakness in someone, something that might cause them to fail, we focus on the weakness until we are sure it will not cause failure. Fixing a weakness is important in preventing failure; however, excellence is never reached by just fixing. We achieve excellence by building on strengths. Our brain grows most where it is already strong.

Tom Landry, legendary coach of the Dallas Cowboys for 19 years, had players review only highlight tapes of their moments of excellence, for example, when they made a great catch, not when they dropped a key touchdown pass. People learn best by building on what they already do best. Michael Jordan would watch his post-game highlights and say of himself, "I did that?" He would then go out and do it again or even better the next game.[3]

The most effective work environments have a lot more "Yes, more of that!" than "Stop that!" While different experts will state different ratios on the healthiest work environments, a ratio of at least four acknowledgments of the positive to one confronting of what is wrong is key to increasing performance.[4] When it comes to a fatal flaw, however, whether in a leader's behavior or with a strong negative perception which is killing their effectiveness, I have seen where a very clear "Stop That and Do Not Ever Do That Again" works magic. As soon as the leader stops a certain behavior, their effectiveness advances.

> "*Excellence is cultivated by building on the best of each person.*"

When it is time to discuss a "stop that" with someone, it is important to be clear, respectful, and helpful in the conversation. Conversations are a two-way flow, allowing the person receiving the improvement feedback to have space to talk.

Consider the following flow in an improvement feedback conversation:

- Clearly state the behavior that you, as the coach, have observed or the situation to be discussed.
- Ask for their point of view. Understand their reality, listen to how they see the behavior or situation, examine their underlying assumptions, look for where they are getting stuck, and discuss what did not work.
- Share your reality, what you think is working, where you see them getting stuck.
- If they improve, take notice and recognize the effort made. If they do not improve, take notice that nothing changed and repeat feedback, coaching.

As a coach, when it comes to feedback, positive or improvement, you are a teacher, always teaching what is important to you and the organization.

- Are you clear on what you want to teach in giving the feedback?
- Are you setting clear expectations?
- Are the goals attainable?
- What are you paying attention to?
- Are you recognizing progress as well as results?
- Are you providing clear feedback and helpful coaching, both always linked?
- Are you holding people accountable for performance?

Remember, there is an emotion attached in working for you. Handling recognition well with individuals and teams can help move the experience of working for you to the positive end of the emotional scale.

If you are leading others today, what feelings do you think others have when they experience working for you?

DAY 25
NEVER SAY "STOP" ONLY

"The chains of habit are generally too small to be felt until they are too strong to be broken."

SAMUEL JOHNSON, 18TH CENTURY WRITER[1]

People do not change that much; but when we do decide to change certain behaviors, we find some behaviors easy to change, others more difficult. Since early in life, our brains have been looking for ways to make our lives easier by forming habits that take little thought. Consider driving a car. Not until the mid-1990s, when I first drove in the United Kingdom, did I learn how little we think when we drive in our normal world. Switching to the "wrong" side of the road elevated all my senses back to the basics of driving. I was no longer driving in automatic mode, for everything was opposite. But soon enough, I began to move away from it being the "wrong" side of the road to being simply the "other" side of the road. A new way of seeing the road became a new normal to drive. Quickly I was driving in automatic mode again without much thought. Changing habits can be easy.

Yet, at other times, changing habits involves moving from a habit initially woven like fabric that has now become a cluster of steel cables, much more difficult to break. In these cases, consider the following when coaching others, or yourself, in making change:

1. Identify the behavior change to be made.

    Be concise. Be specific. Make a specific request for the change. The request cannot be vague or just based on a general dissatisfaction of the person. Also, avoid trying to coach a change in attitude. A person may be negative, rude, lazy, unkind, irresponsible, or undependable–or all of them–but to coach a change, one needs to know what the undesired behavior looks like. Ask yourself, "If I videotaped the person's behavior, what would the person (or I, if trying to change myself) be doing that I would clearly say is negative or rude, or unkind, or irresponsible, or undependable?"

    Write down that behavior in one or two sentences and ask yourself, "Can anyone reading this statement clearly understand the behavioral change that needs to be made?" If your answer is yes, then that is the specific request you make of the person or the behavior target you make for yourself. Coach behavior, not attitude!

2. Never say "No" or "Stop" only.

    When was the last time you tried to change a personal habit that was no longer serving you well? If you are like me, you made the decision to do so, and then you went to work to STOP doing it. We set up reminders to tell us not to do that habit. We fight hard to resist the habit. We tell ourselves, "No! Stop it! Do not do it again!" Honestly, I find that I am not very good at working hard to stop a specific habit. Maybe you share the same experience. Why? Because when we become aware of the habit that needs to be changed, it is already a steel cable, practically impossible to break. If we only try to stop a behavior or say never do it again, the cable may bend, but it will stay in place. We work very hard to stop, but we never break the cable. While a cable cannot be broken, it can be replaced.

"We do not break habits, we replace them."

Putting off a habit, while at the same time putting on a new habit has a much better chance of impacting a change in behavior. Saying, "Start, start, start," or "Yes, yes, yes," is more powerful than saying only, "Stop, stop, stop," or "No, no, no."

Quick example. A supervisor goes into the operations area, always looking for something to fix. Rarely are things running well enough to please him. His conversations with the operators are so negative that, when they see him coming, they often look for places to hide! When given this feedback, the supervisor says, "Ok, I will not go out in the area any longer!" Stop, Stop, Stop! While that might certainly make life more pleasant for the operators, he is the area manager, so staying in his office all day is not effective leadership.

So, what can he "put on" while trying to "put off" his negative interactions with the operators? He starts each day first looking for something going well in the operations area. He would not be doing his job if he ignored things going wrong; but looking first through the lens of what is going well, and having conversations with the operators about improvements and progress, helps to create a more positive work environment. By adding yes, yes, yes (looking for the positive), he is better able to say no, no, no, to the narrow focus on the negative that has impacted his leadership effectiveness.

We do not break habits, we replace them.

"Putting off" a habit is half the battle in making change, but alone it is not likely to create change. "Putting off" while "putting on" is key. Whether looking to change a habit impacting your leadership effectiveness or battling to change a habit that is hindering the quality of your life, never say "No" or "Stop" only.

"No person becomes suddenly different from his habits and cherished thoughts."

JOSHUA CHAMBERLAIN, COLLEGE PROFESSOR, COLONEL IN THE UNION ARMY DURING THE AMERICAN CIVIL WAR[2]

"Put off anger, wrath, malice, and slander. Put on then compassionate hearts, kindness, humility, meekness, and patience, bearing with one another and, if one has a complaint against another, forgiving each other; as the Lord has forgiven you, so you also must forgive. And above all these, put on love, which binds everything together in perfect harmony."

COLOSSIANS 3:8,12-14 (ESV)

DAY 26

BUSYNESS BRINGS DISAPPOINTMENT

Christopher Robin:	What do I like doing best? Nothing.
Winnie-the-Pooh:	How do you do Nothing?
Christopher Robin:	It is when people call out, "What are you going to do?"
	And you say "Nothing," and then you go ahead and do it.
Winnie-the-Pooh:	Ah yes, doing nothing often leads to the very best of something.[1]

Busyness. The dominant daily path chosen by leaders. No, maybe deeper–the dominant daily path chosen for most lives. Why? Because it naturally happens, and it is very easy just to go with the flow. Life is busy, work is busy, and the reflection for today is not a criticism of being engaged and moving through the day to accomplish what needs to be done. The question for the coach, however, is how a routine of daily busyness impacts those who follow you and look to you for coaching.

"Speed is addictive; it undermines nearly everything in life that really matters; quality, compassion, depth, creativity, appreciation, and real relationship."

TONY SCHWARTZ, JOURNALIST AND AUTHOR[2]

I believe the biggest risk of letting busyness be your daily routine is its impact on what we discussed on Day 23, what do you pay attention to? If your day is marked by busyness, when do you lift your head and engage with others during the day? Most likely, when things are not going well or need fixing. Then when the crisis is over, you go back to whatever busyness you needed to attend to, trapped in the "tyranny of the urgent." With this routine, you do not have the opportunity to notice what good others are doing or the value they are adding. Over time your busyness will leave others disappointed, and disappointed team members at some point choose to give less discretionary effort, less commitment to the team, business, or organization. (Remember, people crave attention and more so crave attention on what they do best.)

Let's pause a moment and draw a distinction between the word *busyness* and the word *pace*. When you read them and experience them, are they different? Is a different emotion attached to each of them? To me busyness has a feeling of "empty routine," daily routine not connected to value and worth. Taking what the day gives us. Pace feels different. Pace has the feeling of "choice," control of the speed at which we live our day.

Consider choosing the pace at which you live your day. This does not erase the reality of intense or high demands on your time during a given day or week. Certain demands on your time will cause you to quicken your pace. The challenge lies in what you look for after a period of intensity and high demand on your time. Does the completion of one high-demanding task trigger you to seek more of the same? (Remember, without first becoming aware, you cannot move to improvement.) Increase your awareness of what you seek after completing a high-demanding task. Only then can you better move toward a more sustainable pace.

Choosing pace does not equate to moving slowly. Very few leadership roles allow you to run at a slow pace. Quickness and urgency are the new reality and expectation. Yet a quick pace is different from a hurried, busy pace. Coach John Wooden taught his basketball players, "*Be quick, but don't hurry.*"[3] Without quickness, they would rarely get off a shot, for the defensive player would prevent a good shot. Yet, hurrying to take the shot would rarely result in making the shot because hurriedness impacts form and accuracy. With quickness in your pace, you advance what is important to the task at hand. With hurriedness you stress, think less clearly, make more mistakes, and advance matters far less.

> "We could all benefit from a little
> more nothing."

What are new triggers that can help you move from busyness to pace? We are all creatures of habit, and triggers return us to our habits. Consider adding a trigger for pace by scheduling time into your calendar to slow down. What if your calendar had set times for "nothing"?

- Time when you can lift your head out of your work.
- Time when you can choose to walk out of your office and create a touchpoint with others.
- Time when you can pick up the phone to call someone.
- Time when you can send an email to thank a team member for good work recently done on a project.
- Time when you can simply sit and find value in the stillness.

The list goes on: scheduled time to ensure you connect each day to activities of value and worth. Pre-scheduling time to do "nothing" can help trigger you to connect to individuals, know their stories, understand how they bring value, and express a message of value to them. Give people what they crave (attention), and now you are doing "something" that encourages them to give more of who they are to what they do.

Maybe again, Winnie the Pooh is right: *Doing nothing often leads to the very best kind of something.*[4] We could all benefit from a little more nothing.

"Sometimes when I go somewhere and I wait,
somewhere comes to me."

Winnie-the-Pooh[5]

"I don't believe that good work is ever done in a hurry."

C.S. Lewis[6]

DAY 27

IT'S NOT ABOUT THE QUESTION

"You're not teaching somebody to be a robot and do what you tell them to do. You're teaching a person to think, use his own mind and develop with what you've given him to be able to do it his own way."

JACK NICKLAUS[1]

Questions are indeed an important "friend" to a coach, a friend to keep close by and a key part of a coach's success. On Day 15 we reflected on "Sometimes We Tell, Mostly We Ask." While we certainly want to know how to "tell well," when others need us to, coaches do not reside in the "land of tell," but rather consistently paddle upstream from the habit of telling and instructing to the habit of asking well.

"The best questions are driven from curiosity, seeking what the person already knows."

On Day 14, we reflected on the intention of a coach, the journey of telling-advising-teaching-asking. Questions, within the coaching conversation, are the tool to sharpen focus, increase awareness, create responsibility, and facilitate mobility (FARM) of the person being coached. Questions help coaches suspend their own agenda and focus on the agenda of the person. With curiosity as the motivation behind the questions, coaches use questions as a mirror to help others reflect and better understand what they already know, what matters to them, and where they want to go. The point is not what you think the answers to your questions should be but that others pause and think about their own answers to your questions. Coaches stay on the journey of finding value in the questions they ask, not in the answers they provide, others in the coaching conversation.

So, what makes a good coaching question? I think it helps to consider the mechanics of a good question without making the asking of questions mechanical:

- Most often, they are open ended, cannot be answered with a simple yes or no response.
- Most often, they start (in the English language) with "what" or "how," but "why" questions can also be helpful.
- "What" and "how" questions are forward facing, helping the brain to work more like a google search engine, searching and moving forward.
- "Why" questions are backward facing causing the brain to step back and pause to analyze and justify. Just be careful, as "why" questions can also generate a defensive response if the person feels that seeking blame is the intent.
- The best questions are driven from curiosity, seeking what the person already knows.
- Questions are most helpful when they serve as a mirror to sharpen focus, increase awareness, create responsibility, and facilitate mobility of the person being coached.
- Is it ok to have a list of favorite questions that seem to add value across different coaching conversations? I have some favorites:
  - o What else? (very powerful if asked about three times when seeking options; but ask it ten times, and the person may never have another coaching conversation with you!)
  - o What could you do differently?

o What will better look like?

o If you do nothing, what happens?

o What do you want?

o If the next step does not work, what will you do?

o What advice would you give someone else in your situation?

o If you are saying yes to this, what are you saying no to?

o How can I help?

While I hope an overview of a few mechanics behind good coaching questions is helpful, one of the challenges on my own coaching journey, and maybe also a struggle for you, is to free ourselves from the "fear of not asking the perfect next question." We desire so much to help the person by asking them that one great next question that will help them to move to a better place, that the asking of a question becomes an interference to our coaching effectiveness. What is a more effective focus than what question do we ask next? (Remember, where you focus is where you will go.) Presence! Focus on staying in presence, listening to what the person is saying, staying with curiosity, following their interest in the conversation, and you may be amazed how the next question seems "to lie right in front of you" as the conversation flows.

Practice becoming more comfortable with silence after you have asked a question, allowing the person to pause, think, and respond. Letting the silence sit there, uninterrupted by your voice, is challenging; but if you can get comfortable with five or ten seconds of silence, you will almost always find that the other person will start talking before you can no longer resist the overwhelming urge to say something. Listening for the next question reduces the interference of thinking what perfect question to ask next.

When we give presence to another, we send the message of respect and value. When we ask questions, we are sending the message of "You are competent." And when presence and the asking of questions are joined together, we make people feel visible, seen by another; and that is what all of us desire–to be seen and known by the world around us.

"The confident ask questions to learn what will connect. The insecure just keep talking with the hope something will stick."

SIMON SINEK, AUTHOR & MOTIVATIONAL SPEAKER[2]

DAY 28
USE GROW TO CULTIVATE FARM

"The art of helpful conversation is at risk in our culture. It is so
convenient to take others down in 280 characters or less."

UNKNOWN REFERENCE TO TWITTER
THE MOST COMMON LENGTH OF A TWEET IS 33 CHARACTERS.[1]

We have traveled through many daily reflections with several days focused on foundational coaching concepts and skills such as curiosity in
conversation, presence, FARM, performance interference, coaching intent, coaching lens, feedback, power of attention, replacing habits, and
the use of questions. That is a lot to ponder and focus on; and sometimes
the concepts, as well as the skill building, can be sources of interference
if they take your primary focus off the coaching conversation. The concepts and skills add value only as they serve to increase your comfort and
eventual mastery of the coaching conversation, the pathway to excellence
in coaching.

Today let's reflect on two coaching tools that provide some structure
to the coaching conversation. Coaching tools are like any other tools in a
toolbox. You use them as they are needed to be effective in accomplishing
the task before you, which in this case is generating the product (fruit)
of the coaching conversation with another person: FARM (Sharpening
Focus, Increasing Awareness, Creating Responsibility, Facilitating

Mobility). Some of you may not need the added structure of a coaching tool. But using a more structured tool should not be viewed as a weakness, for what you use in coaching to help another achieve FARM is far less important than the other person achieving FARM! So once again, avoid letting the use of a coaching tool (or not using) become a source of interference in your coaching effectiveness.

One of the two coaching conversation tools that have been helpful to me over the years is the GROW model. Since it is a model or framework, it is most valuable when used in context with other coaching mindsets and skills, which we have reflected on throughout the daily readings. The GROW Model was co-created by Sir John Whitmore and colleagues in the late 1980s to serve as a tool for performance improvement and decision making.[2] It easily connects to FARM, which might be a helpful way to see how using GROW can help you stay focused on the outcome of FARM:

"The concepts and skills add value only as they serve to increase your comfort and eventual mastery of the coaching conversation, the pathway to excellence in coaching."

Goal—What do you want?	Sharpening Focus
Reality—What is happening?	Increasing Awareness
Options—What could you do?	Creating Responsibility
Will—What will you do?	Facilitating Mobility

GROW is most helpful in situations when someone comes to you with a question like, "I have a problem, what should I do?" They can get where they need to go but need a coach to help them think through the best way to get there. Note that you start the conversation with identifying the **G**oal, or as we have referred to it, "the bottom of the slope." Once that is clear to the person, use the coaching conversation to help them think through what has and is happening (**R**eality). Are there interferences keeping them from moving down the slope? Next, the person examines the different ways to get to "the bottom of the slope" (**O**ptions); and finally, they commit to what they will do to start moving to where they want to go, "the bottom of the slope" (**W**ill).

I have found that GROW is a natural conversation style to help the coach keep the person being coached focused and moving. I have listed resources for more information, course work, and even certification in the use and teaching of the GROW model in the Endnotes.

The second coaching conversation tool that has been helpful to me over the years is STATE–WAIT–REMIND–ASK–AGREE–TAKE NOTICE. This tool is most helpful when you notice something that is not the way it needs to be, and you want the person to change. Hopefully the person has a willingness to change, but this tool can be helpful if un-willingness is present or there is a lower level of talent and skills to get to "the bottom of the slope."

STATE	The behavior/situation you have observed.
WAIT	For a response. Be comfortable with silence. Stay in presence. Let silence and performance rest on the person.
REMIND	The person of the proper performance or expectation. Avoid sidetracks and debate. A helpful phrase for keeping on track can be, "Right now I would like to stay focused……"
ASK	For a specific solution. "So, what could you specifically do differently?"
AGREE	On the solution. "What will you do?"
TAKE NOTICE	As the person corrects the behavior/situation. Review and adjust if needed to ensure change/improvement.

The key to success with this tool is keeping the conversation moving, using silence to send the message that this is intended to be a conversation and not a lecture by you the coach, and not letting the conversation get sidetracked or stuck in debate or arguing. In the Endnotes, you will also find a resource for this tool.

Put these tools in your coaching toolbox and use them as you need to cultivate FARM.

DAY 29
THE ACTUAL LIFE WE LIVE EACH DAY

Winnie-the-Pooh:	What day is it?
Christopher Robin:	It's today.
Winnie-the-Pooh:	My favorite day.[1]

Investing in others takes time. While we want to avoid putting coaching and investing in people on a checklist of things to do, we do need to consider how to make it part of our daily routine. Yet, when is a day ever routine? Even if you have made progress in no longer being driven by the mindset of busyness, each day has its own challenges for us to stay focused on the results needed for the day, as well as the increased desire to be in presence with others. We face change every day, and its speed within organizations is on hyper-drive. Our personal lives hear change happening around the world on 24/7 news and social media. But time remains completely unchanged.

"Within limits, we can substitute one resource for another, capital for human labor, for instance. We can use more knowledge or more brawn. But there is no substitute for time. Everything requires time. All work takes place in time and uses up time. Yet people take for granted this unique, irreplaceable, and necessary resource. Nothing else, perhaps, distinguishes effective executives as much as their tender loving care of time."

PETER DRUCKER[2]

I am very comfortable saying the same is true of the most effective coaches. We can always improve on the skills and techniques of coaching, many which we have reflected upon in the daily reflections. Yet, without wrestling with how we view time each day, we will always fall short of having the fullest impact on the individuals, teams, organizations, and families within our world. Each day gives us the opportunity to wrestle with busyness, perspective, and time, shaping how we coach.

The "*tender loving care of time*" (above Drucker quote). We probably need to read that statement every day. C.S. Lewis adds another perspective on time: "*If we only see time through the lens, 'My time is my own,' the more claims we make on our time and the more injury we feel when others interrupt our day. How do we regard the unpleasant things or interruptions in our day? What if we looked at all the unpleasant things or interruptions of the day as precisely our 'real day'? Not nuisances or events with no purpose, but one's real life, our life one day at a time.*"[3]

Take a moment to consider: How do you handle time taken from you? Are you aware? Do you know? What do you consider as interruptions? What are the nuisances of your day? Is time really yours? What if we looked at the interruptions and perceived nuisances of the day as the actual life of that day rather than as disruptions of the day that we had planned? Viewing interruptions and perceived nuisances that way might work magic in the effectiveness of our coaching, but not without some soul searching.

- Are others more important than me?
- Do I think more highly of others then I do myself?
- Am I really interested in others, and do I care about their development and growth in their work and personal lives?
- Is my day so exactly planned that there is no time for interruptions?
- Is there no time for others who need me at a certain moment in the day?

"Each day gives us the opportunity to wrestle with busyness, perspective, and time, shaping how we coach."

Is there any related caution that we need to consider? Potentially, we could start to believe that we must always be available for our people. When we are not constantly available, we feel guilty. When we get too busy, we feel we have failed them. Balance is needed, or we fail to protect the time we need for ourselves. Flight attendants always remind us that should the cabin lose pressure we are to place the oxygen masks over our own mouth and nose before assisting others. Likewise, if we fail to attend to our own needs and make that time a priority, we eventually will not be able to add value to others. There is a season and a time for every matter.

"The Past is frozen and no longer flows. The Present is all lit up. Give patience and gratitude to the present moment."

C.S. Lewis[4]

To Pooh balloons, honey, silliness, friendship, and living in the present, today, were important. Maybe that silly old bear can remind us that today is really the only day we have for sure. If we can focus on the day given to us, TODAY, then maybe, as coaches, we can make today a better foundation for impacting others tomorrow–staying in presence with others, appreciating the moments, and finding gratitude in each day.

"All we have to decide is what to do with the time that is given us."

Gandalf to Frodo, J.R.R. Tolkien, *The Fellowship of the Ring*[5]

"For tomorrow belongs to the people who prepare for it today."

African Proverb[6]

DAY 30
WHERE ARE YOU FOCUSED, AND WHERE WILL YOU GO?

"How did it get so late so soon?"

Dr. Seuss, Author of children's books[1]

We have arrived at daily reflection #30! From the perspective of being a coach, I have benefited from gathering the reflections (lessons learned) from my coaching journey, putting them on paper, and discovering new ways to improve my own coaching abilities. Where we focus is where we go, and with my focus absorbed in writing this book, I know today I help others discover, learn, and grow better than I did when I started writing on Day 1.

But this is a book on coaching, so it is not about my journey. The next step is about you!

"Where are you focused, and where will you go?" With FARM in mind, what would be the best "bottom of the slope" (goal) for you to focus on right now that will increase your awareness of where you are today, create the responsibility (ownership) to move, and then enable you to choose what actions will move you to where you want to go?

Below is a list of challenges, taken from some of the daily reflections, for your consideration. You may already have created your own list as you read through the book. As you reflect on what to focus on, do not choose ten things, six things, or maybe not even two things to work on at the same time. If you choose only one improvement focus for now and move on it, you will be a better coach tomorrow than you are today. Improving your coaching effectiveness is not the time to ask the brain to toggle from one improvement focus to another; rather, it is the time to sharpen your focus and increase awareness on one change you could make now that will enable you to coach others to better performance and success.

As you go through the list, consider the words of Yoda, "*Do or do not. There is no try.*"[2]

1. Picture in your mind asking your team to raise their hands to a height that represents the amount of discretionary effort they choose to bring to their work. What do you see? Partially-raised hands? Fully-raised? More of which one? What is one thing you could do for each team member that would motivate them to increase the amount of discretionary effort they choose to give to their work?

2. You are a prism. You reveal the invisible that is within others. Who in your world needs you to believe that a talent, a possibility, lies within them, and is worth the investment to develop? What will you do to help them discover, learn, and grow?

3. How quickly do you move to diagnostic fixing when coaching others, driven by the habit of being the one that solves the issue of the day? Pay attention to where you tend to go in conversations. What do you notice? What would help you stay with curiosity in conversation a little longer?

4. Is your presence with others already better? (It's been 26 days since we reflected on the topic of presence!) The people you coach would greatly benefit from your increased presence, and the most important people in your life are waiting every day for your presence!

5. What is the one thing this week that you could do to move forward as a coach? Increase your awareness of where you are now,

identify where you want to go (the bottom of the slope), focus on it, and start moving. (FARM)

6. Who under your leadership needs a conversation to sharpen focus, increase awareness, and take responsibility (ownership) to move to a better place? (FARM)

7. Not all coaching conversations are comfortable. Rather than avoid a difficult or challenging conversation, step into the discomfort that may be there and start, even if imperfectly, the conversation. What coaching conversation do you need to have that you have been avoiding?

8. Do you need to recognize, admit, learn, and forget a professional or personal failure so you can remove interference that is impacting your effectiveness at work or at home?

9. Does someone need you to help them push the "delete" button regarding a failure in their professional or personal life? Help them to recognize, admit, learn, and forget.

10. As you look through your coaching lens, who needs you to recognize their talent so they can put it to work? What interferences are keeping them from being successful in the workplace? Do they need a change in focus or an increase in focus?

11. What are specific ways you can help your team be more faithful in showing kindness to one another? Is it an expectation that you have communicated? Typically, what you expect and inspect you get.

12. As you have coaching conversations, what are you noticing in your journey of Telling, Advising, Teaching, Asking? If Telling is needed, do you tell well? Are you enjoying too much the role of advisor or rescuer? At the end of each coaching conversation, who has the ownership for moving forward, you or the person being coached?

13. Do you clearly communicate what you expect in terms of results and behaviors? Do you hold people accountable for what they are

capable of doing and for the core behaviors that are important to the organization?

14. When is the last time you listened to a team member's story? Do you stay *interested* in all your team members, or is the degree of your engagement based on how *interesting* the team member is to you?

15. Does your team need new expectations and conversations as a team on how to make every team member feel accepted and a valued contributor to the team? Do you tolerate any exception to the giving of dignity, worth, and value to every team member?

16. Coaching is not about your journeys, yet your journeys shape the coach you are and allow you to bring perspective to coaching conversations that might actually help a team member see their situation in a little better light. Are you transparent and authentic enough in your connections and conversations to let yourself be known to others?

> **"If you choose only one improvement focus for now and move on it, you will be a better coach tomorrow than you are today."**

17. What are you paying attention to in the performance of others? Take time to notice what you pay attention to. Do you talk with team members? Are they convinced that you not only know who they are as a person, but you also know how they bring value to their work and what they do best?

18. Time. How do you view interruptions in your day? What is important to you, not only at work but in life? Are you trapped in the "tyranny of the urgent"? Or maybe a better question, "Do you thrive in the tyranny of the urgent?" What impact is thriving on the urgent having on your effectiveness as a coach, a person?

Enough of a list from me. What is on your list to improve your coaching effectiveness? Invest in yourself by selecting at least one area to focus on improving this week. When you feel that one area to improve has become part of your coaching skill and brand, select another one.

Where are you focused, and where will you go?

EPILOGUE

I am very grateful that you took the time to read the 30 daily reflections and join me now at the end of the book.

Writing this book has been a journey for me: memories of so many destinations, different cultures, friends who are still connected, friends who have become more of a memory now because of time and distance, 30 years of classes and coaching opportunities, things I did well to help others, and things I could have done better. The loss of presence at home with family during the times of travel is something I think of often, the lost conversations and missed opportunities to make a connection at the right time. It helps for me to remember that *"The Past is frozen and no longer flows. The Present is all lit up."*[1] May I be grateful now for each present moment. I am more aware today that when we say yes to something we are always saying no to something else. Choices are never neutral and always have an impact. The old saying is certainly true: *"The days are long, but the years are short."*[2]

I am very grateful for the opportunity given to me by Eastman Chemical Company and for the opportunities that so many leaders, at all levels within the organization, gave me to walk beside them and learn about leadership and what makes people and work mesh well together. It was a career I could never have imagined. As all of us desire, I hope my journey made a difference for others; and where I did fall short, may I keep those lessons in my heart and apply them each day to be better when with my family and friends and in the coaching of others.

Clearly, you have seen in the book my love for quotes, for things said by others much better than I could ever say them myself. The right quote at the right time has often encouraged me to challenge my perspective and look at my actions. Simply pause for a moment, reflect, and determine if I need to move to a better place. Hopefully, some of the quotes have been an encouragement to you as well, maybe even the words of dear Winnie-the-Pooh, as strange as it may have been for you to find quotes from a "silly ole bear" in a book about coaching! The character of Pooh bear and all his friends from Christopher Robin to Tigger were published following the brutality of World War I and provided a needed solace in a time of sadness, a connection to the innate wonder of childhood.[3] Maybe Winnie-the-Pooh is not just for children and can play a role in helping us gain perspective along our journeys:

Evelyn:	*Do you even like your job?*
(Christopher Robin's wife)	
Christopher Robin:	*What does that have to do with anything? If I work really hard now, then in the future our life will be....*
Evelyn:	*.... Will be what, better, worse? We don't care. We want you. This is life Christopher. This day is your life. Your life is happening now. Right in front of you.*[4]

> **"***It is not necessary to be the best coach on the planet to impact another person.***"**

In the midst of coaching others on their journey to be the best version of themselves, in helping them discover the talents which already lie within, take time to pause with them along the journey and value the growth that is taking place. Brené Brown wonderfully reminds us: *"On the journey for the extraordinary, don't miss life. The ordinary life can bring us much joy, it is not a meaningless life.*"[5] While some people we coach might find greatness, most often people find tremendous satisfaction in the simplicity of day-to-day meaning and purpose.

> *"It is not the extraordinary events that shape our lives. These times are few. Rather, it is the ordinary, day-to-day occurrences that form the pattern of our days and give our lives meaning. It is these events that will someday be read and cherished by those who come after us."*
>
> COL. GARRY A. LITTLETON, FOUNDER, THE GREAT AMERICAN LEATHER STORE[6]

Be faithful and consistent in the simple things you do well, *"Give patience and gratitude to the present moment,"*[7] and when others need a coach, be a good one. It is not necessary to be the best coach on the planet to impact another person. On most days, just making some time for them is the gift they need. You will be what they need, you will be enough.

The satisfaction of a good day comes when we step into the challenges that arise, give our best effort, and bless those around us as we do it. May you always *"Be at your best when your best is needed,"*[8] and as you invest in helping others to accomplish the same, may you experience the blessings that flow from coaching others to discover, learn, and grow.

> *"A life is not important except in the impact it has on other lives."*
>
> JACKIE ROBINSON, FIRST AFRICAN AMERICAN TO PLAY IN MAJOR LEAGUE BASEBALL[9]

> *"Write your name in kindness, love and mercy on the hearts of the thousands you come in contact with year by year, and you will never be forgotten."*
>
> THOMAS CHALMERS[10]

IF YOU FOUND VALUE IN READING MY BOOK....

Please consider giving the book a rating on Amazon or wherever you bought the book. Online book stores are more likely to promote a book when they feel good about its content, and reader reviews are a great barometer for a book's quality. Go to Amazon.com (or wherever you bought the book), search for my name and the book title, and leave a review. If someone gave you a copy of my book, you can leave a review on Amazon. Many thanks in advance for your honest feedback and reviews.

If you know others who would find value in this book, please consider buying a friend, colleague, or family member a copy as a gift. Special bulk discounts are available if you would like your team or other groups to read and benefit from the book. Just contact me at mark@coaching-horizons.net.

If you would like to pursue a coaching engagement with me or schedule a coaching workshop for leaders in your organization, please contact me at mark@coachinghorizons.net to learn more.

Kind Regards,
Mark Hecht

ACKNOWLEDGEMENTS

We all benefit from standing on the shoulders of what others have done and the kindness they extend to us. I have enjoyed writing this section as it has reminded me of so many who made my coaching journey possible. I realize that I will leave out, without intention, others who also blessed me along the way.

I am very grateful,

For Jim Fleshman, Mike Harvey, and Carla Agreda who laid the foundations of my early career and made it easy for me to bring my best to work.

For Edna Kinner who invested in my coaching career, especially my role as an executive coach.

For the gracious hospitality shown to me by so many commercial, technology, functional, manufacturing, and regional leaders, including office and site HR Managers and local HR and administrative support staff for over 25 years. Whether in Asia Pacific, EMEA, or the Americas, I always felt welcomed in the countries and cities that you call home.

For those who directly hosted my journeys across Asia Pacific, EMEA, and the Americas, including those who made the leadership classes possible by handling the logistics of class schedules, participant rosters, and local travel arrangements, some of you for many years: Marian van Kempen, NS Tan, Carlos Panozzo, Rianne van Kampen, Jevgenia Mishkins, Erik de Leeuw, Laura Leiva, Jessie Xu, Helen Chan, Joyce Qi, Estella Zhang, Lydia Yeo, Zuhaira Zubir, Ummi Husna Samsudin, Rachel Ai Phang

Foo, Selina Chan, Cathy Wu, and the always helpful Regional Support Services Team in Rotterdam.

For the Travel and Transport Kingsport team for arranging 2.5 million miles of travel for me, always getting me home, and never leaving me stranded in some distant land.

For the Executive Administrative Assistants, who over many years and with many different Eastman leaders, always found an open time-slot on the calendar of the leaders I needed to meet with: Betty Bailey, Mary Byington, Tammy Cannon, Lynn Drinnon, Teresa East, Darlene Edwards, Becky Flanary, Donna Mosko, Kathy Rhoton, Karen Taylor, and Kala Wallace.

For my last Eastman working team: Russ Brogden, Mitzi Brown, Mike Depollo, Leslie Mann, Mendy Simmers, and Joanne Ward. It was a fun season and we made a difference. Thanks to Gerry Elsea and Scott Kilbourne for the behind the scenes support that made so many classes possible.

For Livia Davis-Jefferies and the Eastman Master Coach Steering Team in giving me the opportunity to work with Joan Peterson, Lisa Stephens, and Jim Boneau from Bluepoint Leadership Development to create the Eastman Chemical Company Master Coach Program. Blessings to the 1st (2016) Master Coach cohort, and all who follow them in the program. As Master Coaches, may you always reveal the invisible in those you coach, helping them to discover, learn, and grow.

For Mark Bogle, Doug Bounds, Bill Fritsch, Bobby Gibbons, Harold Horne, Brad Lich, Tom Morton, TL Ratcliff, Mendy Simmers, Matt Stevens, Peter Briggeman, Erwin Dijkman, Godefroy Motte, Matthijs Veenema, Petra Wood, and Jim Ziegler for their timely support and encouragement during my transition to a new season of life.

For Bonnie, my wife, for reading every word of this book many times and Dr. Beverly White for her professional and diligent role as editor.

For Everett and Malia of Ignite Press and the team they bring together to make a dream-a hope-a book come to reality with hopes of making a difference for those who help others to discover, learn, and grow. I can highly recommend a publishing journey with Ignite Press.

For friends around the globe and here at home who have taught me the art of friendship. Whether friends for over 30 years or new friends within recent years, you have each blessed my life.

For my dad, at the age of 90, and my mom who passed in 2009. For my brother Steve and his family, and my in-laws who are not out-laws, Pop and Grandma.

For my family, Bonnie, Justin, Elizabeth, Brandon, Becky, Lindsey, Amy, Nicholas, Jonathan, Emmalyn, Ryan, Marienne, Addie, Colin, Harrison, Anderson, Charlie, Josalyn, and Ruby, as well as Sam "the Loyal", our faithful Labrador retriever. J.R.R. Tolkien's words, *Not All Who Wander Are Lost*," proved true for me because I always had family to come home to after each long trip.

That I am not my own, but belong to my faithful Savior Jesus Christ. He knows me, calls me redeemed, and will never let me go.

APPENDIX

The Pyramid of Success is a roadmap to successful behaviors. It was developed by Coach John Wooden, who used the Pyramid to train and develop the UCLA men's basketball teams that won 10 NCAA Championships in 12 years (1964-1975). In 1932, Wooden started by creating the "Definition of Success." He continued with the development of the Pyramid to define how to achieve that success. For 14 years Coach Wooden worked on the Pyramid, defining the 14 blocks of the Pyramid below the 15th, which was defined as Competitive Greatness. Later Coach Wooden added "The Mortar," which are 10 blocks that hold the Pyramid together. Today the Pyramid of Success is used by schools, teams, families, and corporations.

I was never able to meet Coach Wooden personally. My first introduction to his values, principles, and life lessons was reading *Wooden: A Lifetime of Observations and Reflections On and Off the Court* by Coach John Wooden with Steve Jamison. From there I fell in love with The Pyramid of Success for its simplicity, yet depth of wisdom.

In the summer of 2017, I attended the first certification class in the John R. Wooden Course. Lynn Guerin, founder and creator of the Wooden course worked personally with Coach Wooden for 15 years to capture the essence of Coach's values, principles, life lessons, and wisdom in the John R. Wooden Course. Lynn not only teaches the certification course each summer at UCLA but is a true coach of the Wooden Way. It was a joy to spend a week soaking in the world of Coach Wooden with Lynn and the participants who attended the first certification class.

Visit the John R. Wooden Course (woodencourse.com) for more information about The Pyramid of Success, the John R. Wooden Course and products related to the values, principles, life lessons, and wisdom of legendary UCLA basketball coach, teacher, author, and philosopher John Robert Wooden.

"Success is peace of mind attained only through self-satisfaction in knowing you made the effort to do the best of which you're capable."

John Wooden

THE PYRAMID OF SUCCESS

COMPETITIVE GREATNESS

POISE · **CONFIDENCE**

CONDITION · **SKILL** · **TEAM SPIRIT**

SELF-CONTROL · **ALERTNESS** · **INITIATIVE** · **INTENTNESS**

INDUSTRIOUSNESS · **FRIENDSHIP** · **LOYALTY** · **COOPERATION** · **ENTHUSIASM**

Faith · Patience · Integrity · Reliability · Honesty · Sincerity

THE JOHN R. WOODEN COURSE

SUCCESS

ENDNOTES AND ADDITIONAL RESOURCES TO HELP YOU GROW AS A COACH

Introduction

1. Maxwell, John C. *Winning With People: Discover the People Principles That Work for You Every Time.* 2004.

Day 1 The Difference a Coach Makes

1. Drucker, Peter, qtd. in Hill, Linda A., and Kent Lineback. *Being the Boss: The 3 Imperatives for Becoming a Great Leader.* 2011.
2. Beck Randall, and Jim Harter. "Managers Account for 70% of Variance in Employee Engagement," *Gallup Business Journal. com.* 2015.
3. Fournier, Camille, qtd. in Berland, Jennifer. "To Friend or Not to Friend: Navigating the Dangers of Social Media at Work." *TrainingIndustry*, 3 June 2019.

Day 2 Giving Hope

1. Michelangelo, qtd. in Flippen, Flip. *The Flip Side: Break Free of the Behaviors That Hold You Back.* 2 May 2007.
2. Certification Program for delivering the Multipliers content. Wiseman, Liz. *Multipliers Revised and Updated: How the Best Leaders Make Everyone Smarter.* 2017.

3. Lewis, C.S. *Letters to Malcolm: Chiefly on Prayer*, letter 6, p. 34. 1973.

4. Bevan, Louise. "Thomas Edison: The mother makes the Genius." *The BL*, 26 February 2019.

5. The John R. Wooden Course, Certification Materials.

6. Koczela, Andrea. "17 Essential (and Authentic) Winnie-the-Pooh Quotes." *BooksTellYouWhy.com*, 18 January 2015. A Disney quote inspired by A.A. Milne, author and creator of Winnie-the-Pooh.

Resources to help you grow as a coach:
Fine, Alan, with Rebecca R. Merrill. *You Already Know How to Be Great: A Simple Way to Remove Interference and Unlock Your Greatest Potential.* 2010.
Wiseman, Liz. *Multipliers, Revised and Updated: How the Best Leaders Make Everyone Smarter.* 2017.

Day 3 Curiosity in Conversation

1. Sales, Jacqueline. "15 Disney Quotes for the Recent College Graduate." *Odyssey*, 4 June 2018.

2. Einstein, Albert, qtd. in Gegelman, Chery. "10 Reasons Why: Curiosity is More Important than Knowledge... Einstein said it. Here's why you need to remember it!" *SimplyUnderstandingOrganizations*.

3. Einstein, Albert, qtd. in Chamorro-Premuzic, Thomas. "Curiosity Is as Important as Intelligence." *HBR*. 27 August 2014.

Day 4 Be Here Now

1. Stoltzfus, Tony. *Leadership Coaching: The Disciplines, Skills, and Heart of a Christian Coach.* 2005.

2. Unknown source.

3. Lee, Dick, and Delmar Hatesohl. "Listening: Our most used communication skill." *Mizzou*, 1993.

4. Cooper, Belle Beth. "10 Surprising Facts About How Our Brains Work." *Buffer.com*. 2016.

5. Thompson, Dargan. "Jim Elliot Quotes That Will Change the

Way Your Think About Sacrifice: On the anniversary of his death, a collection of his greatest sayings." *Relevant Magazine*. 8 January 2016.

6. Whitmore, John. *Coaching for Performance: GROWing People, Performance, and Purpose*, 3rd edition. 2002.

7. Berezow, Alex. "Action bias among elite soccer goalkeepers: The case of penalty kicks." *American Council of Science and Health*. 2018.

8. Brown, Brené. *Dare to Lead: Brave Work. Tough Conversations. Whole Hearts*. 2018.

Day 5 FARM (Focus)

1. Gallwey, W. Timothy. *The Inner Game of Tennis: The Classic Guide to the Mental Side of Peak Performance*. 1974, 2008.

2. Gallo, Carmine. "Steve Jobs: Get Rid of the Crappy Stuff." *Forbes*. 16 May 2011.

Resources to help you grow as a coach:
Gallwey, W. Timothy. *The Inner Game of Tennis: The Classic Guide to the Mental Side of Peak Performance*. 1974, 2008.

Day 6 FARM (Awareness & Responsibility)

1. Quote attributed to Peter Drucker, source unknown.

2. Whitmore, John. *Coaching for Performance: GROWing People, Performance, and Purpose*, 3rd edition. 2002.

3. Gallwey, W. Timothy. *The Inner Game of Work: Focus, Learning, Pleasure, and Mobility in the Workplace*. 2001.

Resources to help you grow as a coach:
Gallwey, W. Timothy. *The Inner Game of Work: Focus, Learning, Pleasure, and Mobility in the Workplace*. 2001.
Whitmore, John. *Coaching for Performance: GROWing People, Performance, and Purpose*, 3rd edition. 2002.

Day 7 FARM (Mobility)

1. Disney. *Christopher Robin*. Motion Picture. 2018.
2. *Online Etymology Dictionary*.
3. Gallwey, W. Timothy. *The Inner Game of Work: Focus, Learning, Pleasure, and Mobility in the Workplace*. 2001.
4. Hecht, Mark. Coaching Horizons.
5. Disney, Walt, qtd. in "The way to get started is to quit talking and begin doing." *Philosiblog*. 22 June 2012.

Resources to help you grow as a coach:
    Gallwey, W. Timothy. *The Inner Game of Work: Focus, Learning, Pleasure, and Mobility in the Workplace*. 2001.
    Whitmore, John. *Coaching for Performance: GROWing People, Performance, and Purpose*, 3rd edition. 2002.

Day 8 Failure & Disappointments

1. Wooden, Coach John, with Steve Jamison. *Wooden: A Lifetime of Observations and Reflections, On and Off the Court*. 1997.
2. Brown, Brené. *Dare to Lead: Brave Work. Tough Conversations. Whole Hearts*. 2018.
3. "24 Winston Churchill Quotes to Inspire You." *Goalpost*, 20 June 2017.

Day 9 The Power to Recognize, Admit, Learn, and Forget

1. The John R. Wooden Course, Certification Materials.
2. Unknown. The analogies of "barnacles on the bottom of a boat" and "failure is like a rip" are not my original words.
3. Nicklaus, Jack, qtd. in Sports Illustrated. "The Golden Rule of the Golden Bear." *Vault*, 21 December 2015.
4. Summitt, Pat, qtd. in The Pat Summitt Foundation. *Facebook. com*. 15 July 2019.
5. Jordan, Michael qtd. in Goldman, Robert, and Stephen Papson. *Nike Culture: The Sign of the Swoosh*, p.49. 1998.

6. Wooden, John. "The Pyramid of Success." Used with permission from The John R. Wooden Course. See Appendix.

7. Sandberg, Sheryl. *Lean In: Women, Work, and the Will to Lead.* 2013.

8. The John R. Wooden Course, Certification Materials.

9. "24 Winston Churchill Quotes to Inspire You To Never Surrender." *Goalpost,* 20 June 2017

Resources to help you grow as a coach:
   Wooden, Coach John, with Steve Jamison. *WOODEN: A Lifetime of Observations and Reflections, On and Off the Court.* 1997.
   Wooden, John, and Jay Carty. *Coach Wooden's Pyramid of Success: Building Blocks for a Better Life.* 2005.

Day 10 Creating Performance

1. Gallwey, W. Timothy. *The Inner Game of Tennis: The Classic Guide to the Mental Side of Peak Performance.* 1974, 2008.

2. Coaching Library. "The State of Workplace Interference." *InsideOutDevelopment,* August 2019.

3. Gallwey, W. Timothy. *The Inner Game of Tennis: The Classic Guide to the Mental Side of Peak Performance.* 1974, 2008.

Resources to help you grow as a coach:
   Fine, Alan, with Rebecca R. Merrill. *You Already Know How to Be Great: A Simple Way to Remove Interference and Unlock Your Greatest Potential.* 2010.
   Gallwey, W. Timothy. *The Inner Game of Tennis: The Classic Guide to the Mental Side of Peak Performance.* 1974, 2008.
   Whitmore, John. *Coaching for Performance: GROWing People, Performance, and Purpose,* 3rd edition. 2002.
   On the Apple or Android App Store, search for TheCoachingApp™ from Profitable Leadership.Designed by Tony Latimer, MCC, BCC.

Search on your internet browser at:
   resouces.insideoutdev.com/articles and search for "The State of Workplace Interference," August 2019

Day 11 Identify, Focus, Remove, Improve

1. Gallwey, W. Timothy. *The Inner Game of Tennis: The Classic Guide to the Mental Side of Peak Performance.* 1974, 2008.
2. Jones, Bobby, qtd. in Aumann, Mark. "Three Reasons Why Bobby Jones Excelled." *PGA.com*, Golf Buzz Series, 18 September 2015.
3. "Henry Ford Quotes." *Brainyquote.com*

Resources to help you grow as a coach:
Tony Latimer has done some excellent work with the Profitable Leadership® that dives further into performance interference and provides a platform for powerful development discussions: On the Apple or Android App Store, search for TheCoachingApp™ from Profitable Leadership. Designed by Tony Latimer, MCC, BCC.

Day 12 Nice is Never Enough

1. Whitmore, John. *Coaching for Performance: GROWing People, Performance, and Purpose,* 3rd edition. 2002.
2. In answer to a question asked by the editors of *Youth*, a journal of Young Israel of Williamsburg, NY. Quoted in *The New York Times*, 20 June 1932, p. 17.

Day 13 But Kindness Is Everything

1. Hadfield, Chris, qtd. in Elton, Chester and Adrian Gostick. "Why some teams don't get along — and what we learned from the commander of the International Space Station." *LinkedIn*, 5 March 2016.
2. Brown, Brené. *Dare to Lead: Brave Work. Tough Conversations. Whole Hearts.* 2018.
3. Saltos, Gabriela Landazuir. "MLK, Jr. asked us 'What are you doing for others?' Here's how we answered." *HuffPost*, 19 January 2015.
4. A modern-day expression most likely crafted from several of John Wesley's sermons of the late 1700s.
5. Edel, Leon. *Henry James: The Master: 1901-1916.* 1972. Also, in

the 2013 February 17, The Chapel Hill News, Section: Chapel Hill News, Column: "My View: Kindness makes a community" by Lynden Harris (Correspondent), these words are attributed to Rogers shortly before he died in 2003: "A great spiritual teacher once said: **There are three ways to ultimate success: The first way is to be kind. The second way is to be kind. The third way is to be kind.**" Thank you, Mr. Rogers, for reminding us.

6. Wooden, Coach John, with Steve Jamison. *Wooden: A Lifetime of Observations and Reflections, On and Off the Court.* 1997.

Day 14 Go Down So Others Can Go Up

1. *The C.S. Lewis Bible.* 2010.
2. Tannenbaum, Robert, and Warren H. Schmidt. "How to Choose a Leadership Pattern." HBR classic. 1958.

Day 15 Sometimes We Tell, Mostly We Ask

1. Lockhart, Keith, qtd. in Greylock Associates. *The Art of Coaching.* DVD. 2002.
2. Gallwey, W. Timothy. *The Inner Game of Tennis: The Classic Guide to the Mental Side of Peak Performance.* 1974, 2008.
3. "Sheryl Sandberg Quotations," page 3. *quotetab.*
4. Peterson, David B. Director, Executive Coaching & Leadership, Google. 2020.
5. Stanier, Michael Bungay. *The Coaching Habit: Say Less, Ask More & Change the Way You Lead Forever.* 2016.
6. Whitmore, John. *Coaching for Performance: GROWing People, Performance, and Purpose,* 3rd edition. 2002.
7. Grahame, Kenneth. *The Wind in the Willows.* 1995.

Resources to help you grow as a coach:
Stanier, Michael Bungay. *The Coaching Habit: Say Less, Ask More & Change the Way You Lead Forever.* 2016.
Stanier, Michael Bungay. *The Advice Trap: Be Humble, Stay Curious and Change the Way You Lead Forever.* 2019.

Search on your internet browser for:
"Michael Bungay Stanier, Be Lazy, Be Curious, Be Often," Transcript & Audio, 2017, at *Blinkist Magazine*.

At YouTube, search for:
"How to Tame Your Advice Monster with Michael Bungay Stanier," TEDx University of Nevada, 2020.

Search on your internet browser for:
"7pathsforward.com and go to publications by David Peterson," Director, Executive Coaching & Leadership at Google.

Day 16 We See What We Seek

1. Disney. *Christopher Robin*. Motion Picture. 2018.
2. Wiseman, Liz. *Multipliers Revised and Updated: How the Best Leaders Make Everyone Smarter*. 2017.

Day 17 Some Things Take No Talent

1. "Women in Sports and Events Atlanta." Facebook.com, 28 June 2016.
2. Wooden, Coach John, with Steve Jamison. *Wooden: A Lifetime of Observations and Reflections, On and Off the Court*. 1997.

Resources to help you grow as a coach:
A great example of TNT applied in the real world is the Miami Dolphins organization. Search on your internet browser for: Brian Flores experience: T.N.T. wall, on-line mantras and more in Miami, by Cameron Wolfe, 2019, at ESPN website.

Day 18 We Don't Make Any Noise

1. Lockhart, Keith, qtd. in Greylock Associates. *The Art of Coaching*. DVD. 2002.
2. Larson, Erik. *The Splendid and the Vile: A Saga of Churchill, Family, and Defiance During the Blitz*. 2020.
3. The John R. Wooden Course, Certification Materials.
4. "The Cost of Leadership: Have You Paid It?" *LeadingWithIntent*.

5. "Positive Choice Integrative Wellness Center." Posting on website. 16 September 2019.
6. Wooden, Coach John, with Steve Jamison. *Wooden: A Lifetime of Observations and Reflections, On and Off the Court.* 1997.
7. Larson, Erik. *The Splendid and the Vile: A Saga of Churchill, Family, and Defiance During the Blitz.* 2020.
8. Wiseman, Liz. *Multipliers, Revised and Updated: How the Best Leaders Make Everyone Smarter.* 2017.
9. Unknown source. Linked most often to Winston Churchill, but no record of when he said it.
10. Glassett, Nick. "27 Mind Expanding Quotes by Lao Tzu." *Origin Leadership*, 2 July 2018.
11. The John R. Wooden Course, Certification Materials.
12. Myers, Courtney Boyd. "The Top 20 Most Inspiring Steve Jobs Quotes." *The Next Web*, 20 Sept. 2011.
13. Wooden, Coach John, with Steve Jamison. *Wooden: A Lifetime of Observations and Reflections, On and Off the Court.* 1997.
14. The John R. Wooden Course, Certification Materials.

Resources to help you grow as a coach:
Wiseman, Liz. *Multipliers, Revised and Updated: How the Best Leaders Make Everyone Smarter.* 2017.

Day 19 Everyone has a Story

1. Unknown. Quote attributed to J.R.R. Tolkien.
2. Bluepoint Leadership Development. Some questions adapted from their coaching workshop.
3. Disney. *Cinderella.* Motion Picture. 2015.
4. Bettinger, Walt, told in Rittenhouse, Laura. "How a Cleaning Lady Inspired Awesome Leadership." *Forbes*, 2 June 2018.
5. The John R. Wooden Course. *Certification Materials.*
6. Warner Brothers. "Flying Home" (Sully's Theme). Clint Eastwood, Christian Jacob, and The Tierney Sutton Band. Released 7 October 2016.

Day 20 Honor Everyone

1. Hesselbein, Frances. *Hesselbein on Leadership*. 2002.
2. LTPI. Bates Communication, Inc.
3. The John R. Wooden Course, Certification Materials.
4. "Anne Frank Quotes." *Brainyquote.*
5. Summitt, Pat. *Reach for the Summit: The Definite Dozen System for Succeeding at Whatever You Do*. 16 March 1998.
6. ESPN.com staff. "The Wizard's Wisdom: 'Woodenisms." 4 June 2010.
7. The John R. Wooden Course, Certification Materials.
8. Meah, Asad. "55 Inspirational African Proverbs on Success." *Awaken the Greatness Within.*
9. Summitt, Pat, with Sally Jenkins. *Sum It Up: A Thousand and Ninety-Eight Victories, a Couple of Irrelevant Losses, and a Life in Perspective*. 2013.

Resources to help you grow as a coach:
Search on your internet browser for:
"Leadership Team Performance Index" at Bates Communication Inc. website.

"Belongingness Is Important to Diversity and Inclusion in the Workplace", by Montrece McNeill Ransom, 2019, at the ABA Journal website.

Day 21 It's About Journeys

1. "Ralph Emerson Henry Quotes." *BrainyQuote.*
2. "Tolkien Best Movie Quotes." *Movie Quotes and More*, 2019.
3. "Allyson Felix Quotes About Running." *quotetab.*
4. Sandberg, Sheryl. *Lean In: Women, Work, and the Will to Lead*. 2013.
5. Lewis, C.S. *Letters to Malcolm: Chiefly on Prayer*, letter 20, p. 106. 1973.
6. The John R. Wooden Course, Certification Materials.
7. Addison, Joseph. *PassItOn.*

Day 22 Feedback Is Not Coaching

1. Soleil, Gina. "Top 10 Principles of Learning & Change." *LinkedIn*, 6 July 2016.
2. Buckingham, Marcus, and Ashley Goodall. *Nine Lies About Work: A Freethinking Leader's Guide to the Real World.* 2019.
3. Lewis, C.S. *Letters to Malcolm: Chiefly on Prayer*, letter 6, p. 34. 1973.
4. Buckingham, Marcus, and Ashley Goodall. *Nine Lies About Work: A Freethinking Leader's Guide to the Real World.* 2019.
5. Kelleher, Herb, qtd. in Greylock Associates. *The Art of Coaching.* DVD. 2002.

Resource to help you grow as a coach:
Buckingham, Marcus, and Ashley Goodall. *Nine Lies About Work: A Freethinking Leader's Guide to the Real World.* 2019.

Day 23 It's About Attention, More Than Feedback

1. Buckingham, Marcus, and Ashley Goodall. *Nine Lies About Work: A Freethinking Leader's Guide to the Real World.* 2019.
2. Study referenced in Buckingham, Marcus, and Ashley Goodall. *Nine Lies About Work: A Freethinking Leader's Guide to the Real World.* 2019.
3. Allan, Leslie. "Gallup Study: Impact of Manager Feedback on Employee Engagement." *Business Performance*, 5 October 2010.
4. Summit, Pat, qtd. in Nia Simone McLoud. "50 Pat Summitt Quotes on Teamwork, Competition, and More." *EverydayPower*, 3 March 2020.

Day 24 Yes, That! Stop That!

1. "Coco Chanel Quotes." *BrainyQuote.*
2. "recognize (v)," *Online Etymology Dictionary.*
3. Jordan, Michael, qtd. In Buckingham, Marcus, and Ashley Goodall. *Nine Lies About Work: A Freethinking Leader's Guide to the Real World.* 2019.

4. Zenger, Jack, and Joseph Folkman. "The Ideal Praise-to-Criticism Ratio." *HBR,* 2013.

Resources to help you grow as a coach:
Brown, Brené. *Dare to Lead: Brave Work. Tough Conversations. Whole Hearts.* 2018.
Buckingham & Goodall. *Nine Lies About Work: A Freethinking Leader's Guide to the Real World.* 2019.
Coyle, Daniel. *The Talent Code: Greatness Isn't Born. It's Grown. Here's How.* 2009.
See HBR article by searching on your internet browser for:
"The Ideal Praise-to-Criticism Ratio," Jack Zenger and Joseph Folkman, 2013.

Day 25 Never Say "Stop" Only

1. Johnson, Samuel, qtd. in Hereford, Z. "Quotes on Habits." *Essential Life Skills.net.*
2. Trulock, Alice Rains. *In the Hands of Providence: Joshua L. Chamberlain and the American Civil War.* 1992.

Day 26 Busyness Brings Disappointment

1. Disney. *Christopher Robin.* Motion Picture. 2018.
2. Schwartz, Tony. "Turning 60: The Twelve Most Important Lessons I've Learned So Far." *Third Culture,* 7 May 2012.
3. Hill, Andrew, with John Wooden. *Be Quick–But Don't Hurry: Finding Success in the Teachings of a Lifetime.* 2001.
4. Disney. *Christopher Robin.* Motion Picture. 2018.
5. Disney. *Christopher Robin.* Motion Picture. 2018.
6. Lewis, C.S. *Letters to Arthur Greeves, The Collected Letters of C.S. Lewis, Volume 1.* 11 July 1916.

Day 27 It's Not About the Question

1. Nicklaus, Jack, qtd. in Greylock Associates. *The Art of Coaching.* DVD. 2002.
2. Sinek, Simon. *AZ Quotes.* 12 March 2010.

Day 28 Use GROW to Cultivate FARM

1. Perez, Sarah. "Twitter's Doubling of Character Count From 140 to 280 Had Little Impact on Length of Tweets." *TechCrunch*, 30 October 2018.
2. PerformanceConsultants.com. Your Global Partner for Coaching and Leadership Development.

Resources to help you grow as a coach:
GROW Model:
Fine, Alan, with Rebecca R. Merrill. *You Already Know How to Be Great: A Simple Way to Remove Interference and Unlock Your Greatest Potential.* 2010.
InsideOutDev.com. *Coaching that works.*
Whitmore, John. *Coaching for Performance: GROWing People, Performance, and Purpose,* 3rd edition. 2002.

State-Wait-Remind-Ask-Agree Process:
Media Partners. *Painless Performance Improvement. DVD.* 1-800-408-5657.
Media Partners. *The Practical Coach 2.* DVD. 1-800-408-5657.

Day 29 The Actual Life We Live Each Day

1. Disney. *Christopher Robin*. Motion Picture. 2018.
2. Drucker, Peter F. *The Essential Drucker: In One Volume, the Best of Sixty Years of Peter Drucker's Essential Writings on Management.* 2001.
3. Lewis, C.S. *The Screwtape Letters.* Letter 21, p. 111-12. 1942; Lewis C.S. *The Collected Letters of C.S. Lewis, volume 2, Letter to Arthur Greeves.* p. 595. 1943.
4. Lewis, C.S. *The Screwtape Letters.* 1942.
5. Tolkien, J.R.R. *The Fellowship of the Ring.* 1954.
6. "African Proverbs." *QuoteWave.*

Day 30 Where Are You Focused and Where Will You Go?

1. "Dr. Seuss Quotes." *BrainyQuote*

2. "The Starwars.com 10: Best Yoda Quotes." *Starwars,* 26 November 2013.

Epilogue

1. Lewis, C.S. *The Screwtape Letters.* 1942.
2. Porter, Janie. "The Best Response to 'The Days Are Long, But the Years are Short." *HuffPost,* 11 April 2016.
3. Sauer, Patrick. "How Winnie-the-Pooh Became a Household Name: The True Story Behind the New Movie, Goodbye Christopher Robin." *Smithsonian Magazine,* 6 November 2017.
4. Disney. *Christopher Robin.* Motion Picture. 2018.
5. Brown, Brené. *Dare to Lead: Brave Work. Tough Conversations. Whole Hearts.* 2018.
6. The Great American Leather Store. Quote on Journal Refills.
7. Lewis, C.S. *The Screwtape Letters.* 1942.
8. Wooden, John. "The Pyramid of Success."
9. "Jackie Robinson Quotes." *BrainyQuote.*
10. "Thomas Chalmers Quotes and Sayings." *inspiringquote.us.*

ABOUT THE AUTHOR

Mark Hecht is a leadership coach and workshop facilitator with three decades of service in the chemical industry. His experiences span multiple cultures and locations in over 20 countries across Asia, Europe, and the Americas. His unique global exposure brings an insightful perspective of cultural diversity and individual uniqueness to his coaching and workshop experiences. As a coach, Mark is recognized for his ability to connect with leaders, investing in their journey to excellence, while also encouraging focus on personal and professional purpose. His coaching workshops focus on the leader's ability to connect with people and create coaching conversations that matter. Whether you are starting a leadership role for the first time or leading at the senior leadership level, Mark can help you learn to draw upon your unique talents to improve your leadership and personal effectiveness.

Since 1981 Mark and his wife Bonnie have lived in East Tennessee, where they raised five children. Today they find tremendous blessings in seeing their children live their lives, whether through the addition of grandchildren, "grand pups", new careers, or new places they call home. For Mark and Bonnie, a growing and active family gives many reasons for hearts of gratitude. New friends have made recent years a special season in life for learning the art of friendship.

In 2016, with 25 years of global coaching stamped in his passport, Mark retired from Eastman Chemical Company and started Coaching Horizons to continue investing in the excellence of leaders. In 2019 he partnered with The Summit Companies in Bristol, TN as a leadership coach to focus on opportunities with local businesses and nonprofits. The Summit Center for Professional Growth was started in 2019 to provide leaders in the local area practical and valuable leadership skills. In addition to his global journeys, Mark has served on the board of Providence Academy, a local Classical Christian School, for 15 years.

Certifications and Education

Gallup Certified Strengths Coach (StrengthsFinder)

Bates Executive Presence Index (ExPI) Certified Coach

Myers Briggs Type Indicator (MBTI) Certification

Inner Game of Performance
(Coaching workshop certification by Tim Gallwey)

Multipliers: How the Best Leaders Make Everyone Smarter
(Course certification by the Wiseman Group)

The Leader Within: A Journey in Self-Awareness
(Leadership Emotional Intelligence course certification by
Bluepoint Leadership Development)

InsideOut Development
(Coaching course certification)

The John R. Wooden Leadership Course
(Course certification)

Member of the Institute of Coaching at McLean Hospital, Harvard

BS in Business Administration, Virginia Tech, 1981

Executive and Leadership Coaching Certificate, UT at Dallas, 2007
(ICF Accredited Coach Training Program)

CPSIA information can be obtained
at www.ICGtesting.com
Printed in the USA
LVHW020419191020
669130LV00031B/839/J

9 781950 710591